Deep Cyborgs Revealed

Nimra

Table of Contents

Introduction

Figure 2 - Rebreather divers after a dive in the Saint-Lawrence (Photo by Chantal Bergevin)

A few anthropologists have studied humans at sea and underwater. During his immersion in the ocean, Stephan Helmreich (2009) referred to the assemblage of Alvin (The researcher's submarine) with the other scientists as a cyborg—a combination of humans and technologies interlocked in a system of communication and control. After showing the picture above to my supervisor and receiving her comment that "there is clearly some resemblance to the cyborg," I began to give greater credence to this theory. It is true that when I dive, a life-support system that includes artificial lungs, a filter and a brain is connected from inside my mouth and merged to my body. In "A Cyborg Manifesto" from 1985, Donna Haraway (2016) stated that cyborgs have much

to do with regeneration. She cited the example of salamanders that can regrow a limb to illustrate

the regrowth of structure and restoration of function. The technical diver, as a cyborg, utilizes his

decompression time underwater as a regeneration moment, a gap in his life where he has space to

breathe. Or perhaps it is those near-miss moments that are regenerative, since they make divers

feel so alive.

From my own extensive experience as a technical diver, scuba diving becomes technical

diving when there is some sort of overhead—whether virtual or physical—that denies the diver

direct access to the surface. With a physical overhead such as within a cave, a mine or a shipwreck,

the ceiling traps the diver underwater, making it impossible to directly reach the surface in case of

an emergency. In a virtual overhead, the decompression requirement forces the diver to stay

underwater to avoid paralysis or death. In all cases, a technical diver does not have the ability to

exit the water at all times, forcing him to solve all problems underwater. Because of the level of

skill and trust required to perform technical dives, it is unusual to do such dives with unknown

divers; tech-divers form small teams that are very difficult to integrate into. In this research, I

address how technical divers in Outaouais, Quebec, practice this risky sport with unforgiving

consequences. Technical diving in Outaouais is often characterized by cold and dark water where

shipwrecks, mines, and caves offer much for divers to explore[1]. For this book, I focus on

technical

---

[1] The Saint Lawrence River is less than a two hours' drive away from Gatineau and is one of the favorite destinations of tech divers due to its great depth and many shipwrecks. Many tech-divers that I discussed with in Quebec believe that in order to do penetration dives and lay lines they needed to fly to Florida or Mexico, where cave systems are abundant. However, the Ottawa River (Rivière des Outaouais in French) has a major cave system that was kept secret by the original explorer's team (so that other divers would not clutter or damage the cave) for many years and still is not common knowledge for divers. Another environment that provide an overhead structures is abandoned mines; mining exploitation in the region left the abused land with submerged holes frozen in time, providing prime spots for divers.

divers who perform trimix dives deeper than 200 feet[2]. For this type of dive, the diver adds helium

to the breathing mixture to reduce the oxygen content as it can quickly become toxic at higher

concentrations. Very few people ever reach the level of trimix diver, and those that do tend to

know each other. I estimate that in Quebec there are currently fewer than 100 active technical

divers at the hypoxic trimix level[3] . In Outaouais there may be around ten active hypoxic trimix

divers—perhaps 25 if the Ottawa region is included[4]. As a trimix-rebreather instructor, I had the

opportunity to certify a few of these divers, most of them young men, many with a military

background. As readers will quickly find out, there is a major gender imbalance in this story. The

absence of female trimix divers is a representation of the current demographic in the sport,

particularly in Outaouais. The same demographic trend is seen in the Army, which has a very close

link to diving as many divers share a military background. Although I do not address the gender

question in this research, I believe it is important to highlight this relevant gender imbalance from

the onset.

Finally, I should mention that technical diving is the topic of this study but also forms part

of its methodology. In this section, I outline the fieldwork I performed in fall 2019 and winter 2020

by following, participating in, and influencing a group of technical divers in Outaouais. My

research is based on multiple ongoing conversations, preparation, diving and decompressing with

several divers. Also, although I am using the term autoethnography to describe my writing style,

it can be somewhat misleading. I have been diving with some of the participants for more than five

---

[2] As opposed to technical diving, recreational diving limit the depth to 130 feet and does not allow the mixing of
helium. Recreational scuba diver must limit their dive's duration to stay within the no decompression limit.
[3] hypoxic trimix is the term used to describe the gas mix when the oxygen content is so low that it cannot support
human life when breathed at the surface.
[4] Recreational scuba diving is a much more popular sport: Lück (2016) estimate is that there were about 22 million
scuba divers in the world in 2014 (Lück, 2016).

years. When I speak, it is an indissociable mix of many of the tech divers that I have frequented over the years. I've embodied their experiences to a point where my voice in this text is a representation of many intertwined perspectives; it goes so far back that it would be impossible to fragment. This enrichment of my experience due to intimate and intense copresence with other divers in risky situations is where I write from. I also took field notes and shot videos throughout the fieldwork to collect sensorial data that was also reviewed with the divers to recount those experiences. I recorded the dives conducted so that we could review and talk through the videos. I also used this time to reflect on my last twenty years of diving.

Scuba Diving as a Method

Technical divers are the main focus of this book; however, scuba diving itself can also be employed as a method. In this section, I discuss my choice of scuba diving as a method and the advantages and disadvantages of such an approach.

It is quite common to use participant observation as a method within anthropology, and since my topic is tech divers, it was logical for me to participate in the diving itself. But diving here is more than solely participant observation. During his dive aboard the Alvin, Helmreich (2007) immersed himself into the ocean bed in order to hear and listen to the soundscape of the ocean. He subsequently argued that transductive ethnography provides people an opportunity to experience a watery immersion, but that it also takes a technical and cultural translation to shape a sub-aquatic soundscape. Although his dive was very different in that he was a passenger of a miniature submarine and he did not have to rely on his underwater skills to survive, his methodology is essential to this research. In his book *Alien Ocean*, he dedicated a chapter to transduction, which is the process that changes sound as it goes from one medium to another. In order to produce his transductive ethnography, Helmreich (2009, p. 225) utilized multiple hydrophones connected to Alvin to capture the sound of the ocean. Those transducers made it possible to immerse himself in the submarine soundscape, and he later claimed that to think transductively is to think about ethnography as a transduction where the ethnographer is a transducer. I will return to the importance of sound for this book and why I paid particular attention to sound within the videos I recorded.

During a dive, all of the human senses are altered, which can make research more complicated. The difficulty of working underwater is related to the water conditions, which can be unpredictable: if divers are bound to a three-hour decompression stop and the weather changes, they may never get out of the water. It is that feeling of defying death and balancing adrenaline that focuses and sharpens the senses. Extreme sports like technical diving are potentially addictive, and withdrawal symptoms can be both physical and psychological (Krivoschekov & Lushnikov, 2012). Vision is reduced, feeling is almost absent due to the thickness of the dry suit, the sense of smell is not effective, and fins act as antennae or cats' whiskers. However, the sense of hearing increases dramatically, as water carries sounds four times as quickly as air. It could make it difficult for humans to locate themselves in space. Scuba diving has been described by Merchant (2011) as a "special sensory experience" since humans are out of their element. When a large vessel's engine is rumbling, the noise can overload the hearing ability of the diver to a point of creating vertigo, a feeling of discomfort causing a loss of balance. For most divers, communicating by voice is not an option. Instead, divers will use a common sign language to converse. However, the usage of sound and ability to hear distinguishes the novice diver from the master. With years of practice, the tech-diver become aware of its surrounding through noise; he is always paying attention to sound, since this is how he will discover that his tank is leaking precious gas from the regulator that has become a natural extension of the body. Focusing on sound is therefore a matter of life and death. A team of rebreather divers will also develop a jargon of ventriloquism to talk underwater. Finally, one of the most important elements in understanding fellow-divers is to read their eyes, which can reveal the emotions they are feeling and whether they have succumbed to panic—a leading cause of death among technical divers.

Video as a Method

In an article published in 2017, Rasmus Rodineliussen maintained that without experiencing the sensorial milieu, it is difficult to understand the choices made by divers underwater. Thus, pairing a researcher's body with a camera seemed like a rather efficient way to do underwater ethnography. I used a camera to collect data and also present some of the finding to the divers and other researchers. As David MacDougall (2004) explained, visual anthropology can take many shapes: some sees it as a research method, while others see it as a field of study, a teaching method, a type of publishing, or a new approach to anthropological knowledge. For the purpose of this project, I employed visual anthropology as a research method.

In his book *Ethnographic Film,* Karl Heider (2006) provides a grid for classifying films based on their degrees of "ethnographicness" that I originally intended to use as a guide when filming. Heider also explains how to use video as a method in such a way as to present whole acts, whole bodies, whole interactions, and whole people to preserve the integrity of the cultural context. Then, he prescribes how to take footage with a camera and how to present it in a film to be categorized as ethnographic. Finally, he explains the principle of holism in ethnography and recommends keeping distortion in a film to a minimum: cinematography tradition should not take precedence over anthropology. Holism can be challenging for a videographer, as Bruno Latour

(2005) has described: the camera becomes an actor itself more than a tool and changes the interaction between the anthropologist and the subject. In the end, I choose not to produce an ethnographic film but instead to use the video capture as a discussion point with the divers. Holism was therefore not so important to me; I did use the camera to document some technique, but I primarily used it to spark interesting discussion with the participants.

I believed it was important to review the videos with the study participants in order to ask the right questions during the discussion. MacDougall (2004) explains that film gives to the anthropologist a mix of concatenation that can help to understand the signification of a ritual that has been written into bodies and motions of the participants. In the scuba diving context, I believe that the video may be well suited to capture and present routines, "rituals," and non-verbal communication. MacDougall (2004) also states that film has the potential to become a medium for exploring social phenomena and expressing anthropological knowledge.

I have been filming underwater since 2011. As a technical diving instructor, I have been scuba diving for the last 20 years. I consider myself an expert in this sport and I have coached many technical divers. I therefore choose the run-and-gun style or "filming as you go," meaning that I did not set up shots but instead followed the action and filmed what seemed interesting at the time. In addition, taking more than two minutes of clean footage in one single clip was difficult as I needed to ensure that I was breathing a proper gas mixture so that I would not pass out due to hypoxia.

The camera was also a great tool for capturing my experience in the limit-environment. As I ventured into the depths of the flooded mine, for example, sound was an important aspect in revealing my state of mind, since with the use of a hydrophone my breathing could be heard along with the reassuring clicking of the solenoid injecting oxygen and reminding me that my rebreather was functioning properly. The discreet "floop-floop" of the mushroom valves also informed me that the gas within the loop was circulating properly and was filtered by the machine to ensure that I stayed alive in this hostile environment. Indeed, sound is what made the video an immerging underwater experience (Helmreich, 2007). As this book is also centered on my own personal experience as a diver and cameraman, these videos provided me with important feedback. In an autoethnographic fashion, I provide my own stories and perspectives on the sport. In short, the sound captured by the video provided a unique transductive autoethnography of a human integrated by technology pushing the limits in challenging dives. Videos have allowed me to re-live those dives and write in an autoethnographic way. The immersive sound of the clips is what allows me to get in the zone, quite critical to writing. After all, most of this book was conceived while doing decompression.

My literature review is modeled after a scuba regulator. I pull on background information on demand, just like the air flow, when it is required to explain my argumentation. Also, most of the literature review is scattered throughout this book just like bursts of bubbles coming out of a scuba regulator. In the next section I introduce a few theoretical approaches and concepts that will be used throughout.

Divers' Becomings

In search of a theory that would explain how deep divers know what they know and how they do what they do, I discovered some researched on a group of indigenous breath-hold deep diver called the Bajau Laut. Recent news articles (Yong, 2018 and Gibbens, 2018) have argued that the Bajau people have genetically adapted to deep free diving, hence explaining how they can hold their breath for longer than the average human. The Bajau Laut (or simply the Bajau) take free diving to the extreme, staying underwater for as long as 13 minutes and at depths of around 200 ft. This group, also called "Sea-Gypsies" or "Sea Nomads," have been a nomadic seafaring group in Southeast Asia that relied primarily on fishing for subsistence. One of their fishing methods is deep free diving using spears, which requires the Bajau to hold their breath for a large amount of time underwater. Recent studies have claimed that there was a genetic mutation and that this ability was most likely related to natural selection. Llardo et al. (2018) claimed that the Bajau live by subsistence based on breath-hold diving and subsequently demonstrated using a comparative genomic study that natural selection has increased their spleen size[5].

I will not debate in this paper the validity of the comparative genomic study, as it is not my field of expertise, nor will I debate whether natural selection can explain the mutation of the Bajau and their capabilities as diver-hunters. However, Nimmo (1968), an anthropologist with significant experience with the Bajau, has criticized several writers who have attributed supernatural feats to the divers in terms of breath-holding capabilities that can reach up to eight minutes. Although he acknowledges that the Bajau do, in fact, possess that impressive capability, he explains that while the Bajau men are able to stay underwater longer than most people, their breath-holding skill is equivalent to other free divers with similar experience. Hence, his observation reinforces that this skill would not be inherited biologically but instead acquired through a lifelong process of practice. This is further validated by Abrahamsson and Schagatay (2014), who mentioned that the Bajau Laut learn to dive at an early age and continue to spend the majority of their life at sea.

---

[5] In fact, it would be the gene PDE10A that would be responsible for providing them with a larger reservoir of oxygenated red blood cells and a strong selection of gene BDKRB2 that would affect the human diving reflex. Those attributes make the genetically modified Bajau superior divers due to their resistance to hypoxia.

Although it seems that some indigenous free divers may have benefited from natural selection, it is impossible to apply this theory to Outaouais tech divers. Because technical diving is a recent activity and the activity is not related to subsistence (at least, in its most basic form), the activity of technical diving is not something practiced by all generations over a long period of time. At most, tech-divers in the region are third-generation divers. In Outaouais there is also no correlation between tech diving and subsistence. As I distanced myself from the nature-culture dichotomy and discarded the neo-Darwinism along with the cultural evolution theory[6], I found that I would not have to bend Ingold's theory of becoming to stick to technical diving. I argue, based on Ingold's theory, that the process of becoming a diver could be a lifelong learning process with ontogenesis at the center. Humans' capacity to perform dives at an incredible depth and duration comes from this process.

In *Biosocial Becomings* (Ingold & Palsson, 2013), Tim Ingold refutes the separation between nature and culture. Ingold outlines the foundation of his theory of evolution thusly: "Evolution, in our view, does not lie in the mutation, recombination, replication, and selection of transmissible traits. It is rather a life process. And at the heart of this process is ontogenesis" (Ingold & Palsson, 2013, p. 9).

In *Evolution and Social Life* (2016 p. xv), Ingold defined ontogenesis as a process of growth and becoming of persons. This process will play an essential part in explaining the

------

[6] In *The Secret of Our Success*, Henrich (2017) explains that cultural evolution is a theory that aims at explaining differences in knowledge essential to subsistence, survival, reproduction, and growth. This theory defines culture as information affecting human behavior that is acquired from members of the same group through teaching, imitation, or other methods of social transmission where language plays a great role In this theory, the concept of culture is defined as ideas, behaviors, and artifacts that can be transmitted and change over time. This knowledge within a group forms a collective brain that no single individual can possess and is therefore shared amongst many. Furthermore, humans are not very good at solving new problems on the spot, but instead rely on generations of compound information packages that are transmitted to new generations

capability of divers to conduct deep dives. I argue that questions regarding deep divers' expertise can be convincingly explained by the process of becoming. To put it in Ingold's words (Ingold, 2016, p. xvi), a diver's becoming "is a developmentally open-ended process of enskilment, as much biological as cultural, rather than the supplemental acquisition of culture upon a universal baseline of human nature." Divers' capability is the output of this process, and it does not make sense to attempt to categorize such output in terms of culture or biology. Essentially, culture is not an information package that can be downloaded over a human body. In *Biosocial Becomings*, Ingold (2013) makes clear that his aim is not to refute that natural selection has existed and continues to shape life on earth or that accumulated knowledge has been transmitted socially by human beings shaping a group collective brain. Rather, it is attribution of the output of a certain capacity to a specific category that is problematic.

In an interview, Ingold warned people about the trap of biosocial becomings (Angosto Ferrández, 2013). If we use the concept lightly and try to bridge the gap between nature and culture, trying to explain certain ethnic groups' capabilities using an argument of biology, for example, this could look like racism. Perhaps after a period like World War II, there was no appetite for using biology to distinguish between different ethnic groups' capabilities. As researchers, we cannot say that differences in cultures are biological if biological is defined as genetic and hereditary. In the same interview, Ingold explained one of his arguments that lays the foundation of his theory:

> "Our humanity is not something that comes with the territory. It's not given by species membership or by belonging to this culture or that. Rather, we have continually to be creating our humanity for ourselves in what we do."

Because of the process of ontogenesis and because our humanity is the result of what we do, the use of biology in anthropology should not be seen as racism. After all, if I had practised breath-hold diving and fishing throughout my life, I could have achieved similar success as the Bajau divers—but perhaps in a different environment. Pitrou (2015), inspired by Ingold, asserted that evolution should not be seen in terms of lineage but instead as mutual relations that are part of a developmental system that shapes forms of life. The usage selection of "forms of life" referring to the "cultural, social, symbolic ways of acting and thinking about human groups" is particularly different from "life forms" which refers to any organism (Helmreich, 2009). In this sense, a deep diver could be a form of life.

Edgework and Limit-Experience

It was in the news in 2019 that many people died on the summit of Mt. Everest, where there seems to be a traffic jam causing many accidents. K2 mountain has a death rate of approximately 29% (Leonard, 2013)[7]. The question that was relevant to this research and that I asked myself was "what is it about climbing Everest that is worth risking your life for?" Because there seems to be little coherence into doing such action, I started to question myself about taking risk voluntarily. Through ethnographic fieldwork with skydivers, Lyng (1990) developed the concept of edgework as a form of volunteer risk-taking in leisure activities with great danger, such as base jumping, waterfall kayaking, scuba diving, or free climbing. People participating in edgework come as close to the line (or edge) between the success and failure of a challenge as possible without crossing it, involving a high amount of skill in the process. Those individuals are not reckless or irresponsible,

---

[7] Whether or not this is accurate is not important, the point is that the odds of dying are significant.

Lyng argues, but instead they understand the risk they are taking and they seek to master the situation.

It seems that edgework is particularly popular among North Americans and in countries with high incomes, where people seek thrills in order to learn about themselves. Edgework does not include activities like bungee jumping, where the aim is to confront fear and anxiety without any skill involved. Commercialization of other sports such as white-water rafting eliminates most of the danger related to skill and therefore does not qualify as edgework. Lyng (1990, p. 851) explains that the concept of edgework "highlight[s] the most sociologically relevant features of voluntary risk taking." His theory emerged from the fact that he was unable to reconcile the pre-existing body of research conducted primarily by psychologists who aimed at understanding what made risk acceptable; the main shortcoming of psychology's personality predisposition models was that it did not explain voluntary risk-taking[8].

While North American society works hard to reduce risks of injury and death, the high-risk sports industry has been growing at a fast pace; Lyng therefore poses the question of why risk-taking is necessary for well-being. As a jump pilot for a group of skydivers, Lyng produced an ethnographic study of the sport and analyzed edgework within a social frame (Lyng & Snow, 1986). Edgework is about negotiating the boundary between chaos and order or death and life, and in order for an activity to qualify it needs to have a clear threat to the participant—that is, failure usually means that the participant will die or be severely injured. Edgework can represent trades such as the infantry, wildfire fighters, or test pilots. The edge applies to the participant but also to

---

[8] The intrinsic motivation model portrayed stress-seeking as stimulation with addictive qualities and self-destructing behavior as a means to counter depression (Farberow, 1980). Lyng (1990) pointed out that this model does not explain the relationship between the psychological factors and the social context of the risk-taking.

the technology, as in the example of a race car driver who attempts to take the vehicle to the limit of traction and speed.

Another theory that share some similarities with edgework is limit-experience. In an interview with Duccio Trombadori, Foucault explains his idea of the limit-experience. He distinguishes Nietzsche, Bataille, and Blanchot from the traditional phenomenologist who would traditionally study the daily experience. Through experience, those three intellectuals would try "to reach that point of life which lies as close as possible to the impossibility of living, which lies at the limit or extreme. They attempt to gather the maximum amount of intensity and impossibility at the same time." (Foucault, 1991, p. 30). Furthermore, Foucault explains that according to them, experience has the duty of " "tearing" the subject from itself in such a way that it is no longer the subject as such, or that it is completely "other" than itself so that it may arrive at its annihilation, its dissociation." (Foucault, 1991, p. 31). For Foucault, it would be this dissociation from self, the "tearing", that would make the idea of limit-experience, which would be an experience that change its subject forever. Much of technical diving has to do with negotiating the edge and limit-experience; the stories from chapter 2 and 3 are great example of it.

Research Question

Divers plan dives in a way that makes sense to them (Lock, 2019, p. 7) . Often novice tech divers will rely on the information available to them at the time of planning the dive; such information is typically comprised of some newly developed skills and a set of universal rules that has been learned in a series of short classes. Most likely, with their limited experience, they have never suffered from the bends. As they progress and increase the difficulty of their dives, they increase the likelihood of bad luck and decompression accidents. While many thinks that by staying within the rules of technical diving one will not suffer from a decompression accident, some quickly find out that it is not the case and sadly they are ill prepared for such event. With a lack of public information on deep diving[9], some will begin to bend the rules to reach depth and stay longer underwater pushing their own limits. Sometimes they will go too far and encounter a near miss event that provides them with a series of mixed feelings. Some divers die. With a clear lack of literature on the subject, I asked myself how one becomes a deep diver, going to depth beyond what is advertised in any training agencies.

The following story is an account of how I became a deep diver. This is by no means a complete diving autobiography, as it would be impossible to document completely the last twenty years of my life. Instead, I have chosen to describe particular events that define how I learned the sport and grew as a diver. As such, I recognize that there may be some aspects in those events that I have left out. The first chapters follow a chronological order and explain how I became a tech diver, beginning with what I call the school of diving—the somewhat formal classes that various agencies provide for students to learn the basics of diving. Subsequently, it describes how I questioned the official knowledge of the sport regarding rules and decompression. The second chapter is a phase of growth beyond the boundaries set in place by the industry; it explains how I have adjusted tech diving to cold weather. The third chapter tells the story of the Minex Project, a microbiology study and exploration of a flooded mine to extreme depths which constitutes the heart of the book. Finally, the fourth chapter unfolds those inevitable near-miss moments that are found to be regenerative.

------

[9] Deep dives occur beyond the limits of technical diving courses currently set at a maximum of 400 feet. (my definition).

Throughout my diving career, I have had the opportunity to meet incredible instructors and mentors who taught me many things that have allowed me to survive all of my dives. None of my accomplishments such as world records belongs to me; they belong to a team. Without the team, none of this would have been possible, and although I could have written hundreds of pages about each of the team members. I have chosen to grant them privacy however I would not be diving if I were alone. If there is only one thing to learn from my entire journey, it is that technical diving is about a tight set of links between very few individuals who share a passion and constantly defy death itself. It is from reaching these depths in many senses of the term that regeneration or regained vitality is found.

If one feels disoriented during the reading of this book, then one needs to hold his breath, close his eyes, and imagine himself in a deep cave. The CO_2 will raise slowly to make one feel the urge to breath, followed by anxiety – one should fight this urge for another 30 second. This way, one may gain a small insight of what a near-mist event is. One should simply keep reading to find the line back to the exit.

Chapter 1. Diving In

Disclaimer

The following text is an ethnography documenting some exploration projects along with the people, technologies and process involved. The diving techniques described in this ethnography, both in theory and in practice, are often outside accepted norms. These diving techniques are not standard procedures and should not be followed. To learn proper diving methodology, I recommend readers take a class with an accredited agency. Also, I never received

any equipment sponsorships. The mention of the equipment's brand is purely for documentation

purposes.

The School of Diving

My passion for diving was kindled at a very early age. As a child I played in water all day

long, developing good swimming skills. When I turned 12, I completed my first scuba diving

session in a pool where I learned the first rule of diving: "never ever hold your breath"[10]. I was

already attracted to the sport growing up, having seen my grandfather's scuba diving equipment

that was made from a homemade pipe folded in a rack's shape that had a strap to attach an old tank

with a J-valve[11] and a single regulator. My father and uncle were also divers. My dad took his

course at the Royal Military College and allowed me to take my open water certification at the age

of 13 in the Dominican Republic; I spent the entire trip studying for the final exam.

---

[10] By holding their breath while ascending, divers could suffer pneumothorax.
[11] Old type of valve that released a reserve of gas when actioned. No longer in use by the majority of divers since the invention of the pressure gage.

Figure 3 - Learning to dive with my father in Dominican Republic (Photo by Anonymous)

After graduating from high school and earning a little bit of money, I was able to buy my first equipment on eBay for about $350 USD. It was old rental gear from California and was already worn out by students and salt water. Nevertheless, I ended up using it for many years, and it is still functioning 15 years later with only basic maintenance I have done myself.

As a young adult, one of my goals was to become a scuba diving instructor, as I wanted to share my passion with others. A very professional retired British officer named Liam was largely responsible for my development as a divemaster through his mentorship when I was trying to become a recreational diving instructor. In order to pass the two-day-long open water diver instructor exam, I drove with a friend from Kingston, Ontario, to Florida and back—all within a five-day period.

In university, I was introduced to the world of nitrox and technical diving by Benjamin, one of the instructors at the local club in Kingston. Nitrox, also called enriched air, is made by adding oxygen to the air to diminish the nitrogen percentage in the cylinder, hence reducing the decompression requirement. With nitrox a diver can spend more time at depth than with air. The notion of nitrox was still somewhat taboo[12] at that time and was not allowed at the club, so the blending of the gas was done directly at Benjamin's house, where he would have a large T-tank (a large bank of gas) of oxygen strapped to the fence in his backyard. Today nitrox is considered "safer than air" because of the reduced decompression requirements. However, it comes with some drawbacks: augmenting the fraction of oxygen in the mix also potentially increases the risk of oxygen toxicity with depth, which leads to convulsions and drowning. This is why mixed gas must always be analyzed and properly identified on the tank itself. Many accidents in diving happen because a technical diver used the wrong gas at the wrong depth, exceeding the partial pressure (PP) limit of oxygen (O_2) for the extended period of time that the human body can tolerate (The Central Nervous System clock exposure table estimates the safe limit of tolerance (Dreher, 2011))[13].

In 2006, at age 19, I completed the advance nitrox and decompression procedure courses. Breathing pressurized gas has been known to be a problem since the end of the 19th century. In fact, French physiologist Paul Bert was the first to successfully identify decompression sickness (DCS) and its cause: breathing air while being under pressure, at depth, would form nitrogen

---

[12] While being legal and commonly used in Florida, nitrox was still not allowed in the local diving club. Due to the lack of knowledge around mixing gas and the danger of compress oxygen made people stay away from such practice.
[13] The partial pressure of gas can be found by multiplying the ambient pressure by the percentage of gas in the mix. For example, at 33 ft of depth (or two atmospheres of pressure), breathing air which contains approximately 21% oxygen would result in a PPO_2 of 0.42. At a PPO_2 of 1.6, the exposure is limited to 45 minutes only, based on most central nervous system (CSN) clock tables.

bubbles in blood and tissues after returning to the surface. But in 1908, a Scottish physiologist named John Scott Haldane published his work on a decompression model for the Royal Navy along with the first decompression tables noting that the human body could tolerate some pressure changes. He established that the maximum pressure ratio a body could tolerate was 1.58. Later, Robert Workman, while working for the U.S. Navy, successfully argued that various tissue types would have different speeds of gas dissolving. For example, bones would take more time to "on gas" (when gas dissolve in the tissue by increasing pressure) and "off gas" (when gas dissolve out of the tissue into the blood stream by decreasing pressure) than muscles, creating various tissue categories for decompression. The most popular decompression model used today, the ZH-L16, was developed by Dr. Albert Buhlmann at the end of the 20th century[14].

For recreational divers, time at depth is limited so that their bottom time never exceeds the no-decompression limit (NDL) prescribed by the body tolerance pressure ratio. For technical divers who break the NDL, a virtual ceiling suggested by the same tolerance ratio appears and limits the ascent. This maximum tolerance ratio, also called M-Value, was tested on a few divers to ensure an acceptable threshold for symptoms of decompression sickness. To perform decompression, divers are required to ascend slightly above the ambient pressure depth, keeping their body in equilibrium with the environment to purge inert gas (helium and nitrogen), but must never pass the ceiling, which is gradually pushed up as tissues off-gas. The margin between those decompression stops and the theoretical maximum value a diver can ascend without risking the bends (or the ceiling) is called the gradient factor (GF) and is represented by two percentages. The higher GF represents how close to the maximum body tolerance pressure ratio the diver will exit

---

[14] The ZH-L16 categorizes tissues into 16 compartments.

the water. For example, with a high GF of 80%, a diver would exit the water when his tissues have dropped to 80% of the M-Value; if he exited the water while still above 100%, the diver would in theory risk the bends (although variations in humans' physiology make this a grey zone). The other percentage is the low GF, which dictates how deep the diver begins his stop as well as indicating the rate of ascent. With a low GF of 30%, a diver would begin his deepest stop when his body reaches 30% of the M-Value.

This linear relationship between M-Value and ambient pressure makes it somewhat easy for a dive computer to predict a decompression schedule. Today, popular dive computers such as the Shearwater Petrel come with a setting of low GF 30% and high GF 70%, also written 30/70, to provide a safe decompression schedule. At the time of completion of my decompression class, I used the very popular US Navy table, a watch, and the "plan your dive and dive your plan" mentality. I believe that my instructor, Benjamin, was very adamant about the Doing it Right (DIR) approach from the Global Underwater Explorers agency (GUE), and even as a Technical Diving International (TDI) instructor, his methods were considered extremely strict by other divers. Aligned with the DIR mindset, he was a competent cave diver diving with an armadillo harness, which I believe was the only commercially available sidemount rig at the time. This is what initiated my curiosity about cave diving. Sidemount means attaching tanks on the diver's side instead of on his back, making him slimmer and allowing him to slip through low-profile caves. For this class, though, I was taught to use the Hogarthian standard regulator configuration with twin tanks. My first set of tanks was a twin low pressure 125, which were sometimes called "hot water tanks" due to their large size—I could literally kick the bottom of my tank with my heels.

Once I was qualified in decompression procedures and was teaching as an open water scuba instructor, I told myself: "That's it, I'm happy with what I'm doing—I don't need to take more classes to go deeper. My depth limit will be 180 ft." I had put some thought into learning to dive rebreather, but the cost and the then-bad reputation of the machine made me think that staying open circuit was the best choice. A closed-circuit rebreather (CCR) is a machine that recycles the breathing gas by removing the carbon dioxide and re-injecting the consumed oxygen. It is chiefly comprised of a set of counter lungs, a carbon dioxide filter, and a sensor/computer that analyzes the gas. A rebreather allows the diver much greater autonomy underwater, permitting dives lasting many hours and removing the dependence on the volume of open circuit tanks. It is much more laborious and task intensive to manage a rebreather than a set of twin tanks—so much so that many divers have passed away on those machines due to operator errors and equipment malfunctions. However, there is a natural point where a dive becomes safer by using rebreather over an open circuit due to extreme depth, duration, or overhead environment.

Upon graduating from university in 2009, I finally had the time to drive to Florida and began my cave diving class with a retired Canadian military captain named Jacob. He did not take it easy on me, but I developed a passion for caves instantly. For the next seven years, I spent most of my vacations diving the caves of Florida and Mexico. It was during this time that I began to feel that my twin tanks were inadequate for the dives I was executing. I wanted to get further and deeper in the caves. I then decided to learn sidemount, which at the time did not have many commercial options, so I decided to build my own rig. It was while sidemounting into a cave that I experienced my first near-miss event: I was completely stuck between two rocks in a zero-visibility exit dealing with a loose line. I will explain this near miss in a later chapter. Although I successfully got out of this situation, it made me rethink my position on open circuit vs. closed circuit. The open circuit is extremely limited for this kind of diving due to the overhead and the rule of thirds, which states that a diver should only use one third of his gas to penetrate into the cave, keeping the second third for the exit and the last third to manage emergencies, such as a partner running out of air. Essentially, an open circuit system seemed like sitting on a time bomb for long cave dives—at one point you will run out of time.

In 2011, after meeting two Quebecers in Ginnie Spring that were diving closed circuit, I decided to try my first rebreather, and not long after I bought my first machine, a Megalodon Copis. The "Meg" had a reputation of being built like a tank, being very reliable, and having the best rate of breathing on the market. As opposed to a fixed percentage of gases in a cylinder, a rebreather requires the diver to mix oxygen on the fly to optimise his decompression schedule; too little or too much oxygen could result in death.

Figure 4 - Diver's donning a Meg rebreather in James Bay (Photo by Kevin Brown)

After gaining much experience on the rebreather, I learned to dive deeper using helium to a limit of 330 ft. I earned my trimix certification with Julie in 2013 in the Floridian and Mexican caves, a passionate instructor, did more than just "qualify" me. She also influenced my development not only on the technical level, but also in the way I am reflexive and reflective about diving. She continued to mentor me beyond the official course and was generous enough to share her experiences regarding some limitations of diving. I now pass those experiences through my own mentoring when I teach novice divers. Trimix is the name used to describe a mix of nitrogen, oxygen, and helium. The primary function of the helium is to be an inert filler, and by replacing nitrogen, helium diminishes nitrogen's narcotic effect at depth (and perhaps oxygen to some extent). Helium also reduces the amount of oxygen in the tank for deeper dives, so that the partial pressure of oxygen remains tolerable for the body. Even air, with 21% oxygen, will become too concentrated for proper utilization at depth. When we "cut" a mix below the point of sustaining human life on the surface, such as by diluting oxygen to 10%, it is deemed hypoxic. For example, breathing a mix of 10% oxygen, 60% helium and 30% nitrogen at the surface will make a diver unconscious and drown.

Although not all training agencies agree on standards, it is generally accepted that a diver needs to keep the oxygen at a partial pressure between 0.18 and 1.6 and the nitrogen below 4.0 to limit the narcotic element. Nitrogen narcosis is quite dangerous and can lead divers to do incomprehensible things and eventually pass out. It is often compared to the feeling of being intoxicated by alcohol. In the 2000s, nitrox, trimix and cave diving were well developed and accepted in Florida, the center of scuba diving. For some reason, Canada (and even more so Quebec) is always far behind in terms of diving techniques and developments. Perhaps this is due to our short warm season or the high price of equipment in dive shops. Trimix is often seen as a complex business, but as a great friend of mine said to summarize its procedure: "l'hélium tu le criss dans ta tank et ça finit là" ("Helium, you put it in your tank and that's it"). Having reached the level of rebreather-trimix-cave diver, I finally felt like I had completed the school of diving: there were no more required certifications—I was a qualified diver.

After the School of Diving

Questioning the Rules

After I completed the school of diving, I started to accumulate a significant amount of experience on my rebreather, some years completing over 300 hours of dive time, many of which were on trimix in caves or shipwrecks. However, while observing more experienced divers, I wondered how they were able to achieve some of their exploration dives. I was already diving at the limit of the "rules" in the books and could not stay as deep or as long as them. I started to realize that some of the great explorers were not following those rules at all. In contrast to what some would say in public or teach their students, they would bend the rules for their own diving. I then started to gradually push the limit of the accepted norms of tech diving little by little as experience began to supersede technical knowledge. To me it was simple, I either had to bend the rules or I would not be able to go explore what I wanted to explore. After all, those tables and rules were built based on human testing on navy divers, and year after year the industry made them stricter by increasing the safety margin.

Pushing the limit of diving was not without consequences. Before the age of 30, I suffered at least 20 decompression accidents "deemed type 1 (a less severe, non-neurological DCS)." I consulted many doctors and specialists on the matter, since for many of those hits I was still following proper decompression schedules; so those bends, in theory, should not have happened. I passed two patent foramen ovale (PFO) tests, the first the normal echocardiogram test and the second the bubble test, which consisted of injecting a solution with small bubbles into my veins and observing their trajectories in my heart. The PFO is a small orifice between the left side and the right side of the heart that should close itself after birth. Although there is a lack of scientific evidence due to testing and statistical reasons, the tech diving community firmly believes that people with PFO are much more susceptible to the bends, since the PFO provides a bypass for the lungs, the natural decompression filter. I have heard that approximately a quarter of the general population have a PFO. Many of my close diving friends discovered their PFOs by suffering from a decompression hit. All had the minor heart surgery to have the hole closed, which involves applying a Gore-Tex patch that looks like an umbrella to the hole. Both results of my PFO tests were negative, but the presence of a small one remains a possibility. There is a third test that is more efficient called a transesophageal echocardiogram, which consists in swallowing a probe to scan the heart, but because I live in Quebec it is difficult to get this test.

Because of that series of bends I had begun to question everything about my diving technique and came to the conclusion that much of my knowledge came from a different environment then the Canadian one and was probably not adapted to my region. My diving environment—my backyard—is the mines, caves, and shipwrecks surrounding the Outaouais region. In the Saint Lawrence River near Brockville, Ontario, we find early shipwrecks from the exploration days as well as more recent cargo transports. In July and August, the water temperature can reach 75° F, although the warm water in the region is correlated with notoriously bad visibility. In the winter ice covers the river, making most of the year cold-water diving conditions. In addition, current in the Thousand Islands near Brockville can surpass four knots when the river squeezes between islands, and knowing how to read the current is essential to a diver's safety.

Many of my dives were requiring over three hours of decompression in cold water and in extremely strong currents. The most exceptional dives required over eight hours of decompression and breaking the CNS clock exposure limit by 400%. After meeting with a researcher at the Canadian Forces research diving unit, I realized that most of the major dives I performed were outside the table of tests in sport and military diving. This meant that I was the test subject for those decompression models. The reason I met with the research diving unit while in the military was that I suffered my first type two decompression hit, a neurological one, on the Roy A. Jodrey shipwreck.

Lying 240 ft deep in the American's Narrow of the Saint Lawrence River, the Jodrey is the deepest-known large shipwreck in the region. It has a reputation for severely bending many divers, including myself. In 2015, after a long dive to the engine room of the Jodrey, it hit me. I was only out of the water for about 15 minutes when I started to have difficulty breathing. It felt like I had fallen on my back and lost my breath; the burning sensation was intense, similar to eating too many jalapenos. I forced pure oxygen into my lungs and slowly my breathing came back, but my skin around my abdomen was turning blue and purple. I made my way back to Gatineau and lay down breathing oxygen while the pain took over my left arm. The next day I went to the military hospital and they immediately transferred me to a hyperbaric chamber where I started a TT6[15] decompression profile for six hours. After about five hours of decompression on pure oxygen and accumulating roughly 1500% of the CNS clock, I felt nauseated. Certain that the convulsions were

---

[15] TT6 decompression profile is usually chosen by the physician for deep diving accident.

near, I signalled the nurse to flush the chamber, giving me an air break. Luckily, there was no permanent damage.

In the months following my decompression accident, I started to look for answers as to why it had happened. I reanalyzed the dive completely: it was well planned and well executed, so what went wrong? My decompression schedule was adequate and included a good safety margin. The water was relatively warm, so cold water could not be the culprit. Because I was in the military at the time, a team of hyperbaric physicians analyzed my situation and the data from my dive computer. Their conclusion was unanimous: you should never dive again. But quitting diving was out of the question. After all, I thought, what do they know? I told myself that those physicians did not dive to the level at which I was diving[16]. I knew one thing, that what got me to bend was the unofficial factors and details, the things one cannot learn in a book but instead must learn by experience—things like the effect of a strong current, hydration levels, physical exhaustion, the type of food I consumed, and how much sleep I had before a dive. This was about a human's capabilities in a borderline environment.

The Cold Water

After beginning to ask hard questions during the period after that series of bends, I started to change the way I dove. My diving style evolved into something new, somewhat less esthetic and more practical for the cold Canadian waters, sacrificing streamlining of gear for accessibility. First I decided to analyze the gradient factor that I was using in a few scenarios. For a five- or six-hour dive at Eagle's Nest in Florida, a well-known deep cave, I was typically using a 10/90 GF

---

[16] Although I recognize that there are physicians that are formidable technical divers.

without any DCS problems. Although not providing a great safety margin, it was quite a common profile at the time. I understand that this is probably considered old school, especially with regard to the mentality of deep stops and the 10% low gradient, but this was before the more recent studies promoting shallower decompression appeared. However, back in Canada in the cold water of the Saint Lawrence River, I experienced small bends using gradients that were considered very conservative: 30/70. If I can afford to add more safety, I usually do.

Other than getting the bends, cold water was and still is my worst nightmare as it drastically increases how risky a dive is. On a four-hour dive that includes a three-hour decompression, a cut in a dry suit in a 38-degree water forces a diver to make some serious decisions. Do you skip decompression and most likely bend to the point of non-return and die? How long can you last in the water before hypothermia makes it completely impossible to control your rebreather? Based on many experiences including a complete flooding of my suit in the Bay of Fundy, I currently think that with a total suit failure I would have roughly one hour before I would be forced to exit the water. That could leave me with two hours of skipped decompression, which is most likely fatal. But not all suit failures are equal—some are severe, others mild.

Although my fitness level is not as good as it was when I was in the military during my twenties, I still work out regularly. I did complete the Petawawa military Ironman and ran a few half marathons. I was not overweight and had a descent cardiovascular capacity, so I do not think that physical fitness was a factor in my decompression accidents. I did not smoke but was a social drinker. Two days before a big trimix dive, I would not drink any alcohol; instead, I would start hydrating with Gatorade.

After many failed attempts at finding the right gear in the constant fight to stay warm, I found motorcycle underwear heated by a radiant cable and decided to modify its wiring to fit on a diving lamp's battery canister. This solution has worked well for me and I still dive with it today. That way, it would keep working even if the suit flooded. At the end of the day divers have a serious risk to acknowledge in cold-water decompression diving. A small cut in the suit would usually be a bad day, which could mean a decompression chamber ride. Yes, I could mitigate some of the cold-water risks by, for example, carrying a second pair of gloves to pull over the first one if it ripped or wearing a tighter neoprene dry suit to minimize the heat loss in case of a flood. But cold-water tech divers put their lives on the line in dangerous environments. Some of those environments, like shipwrecks or mines, may be unexplored, and therefore no prior information is available on their stability or the presence of sharp objects that threaten to pierce a diver's suit.

The Cost of Technical Diving

To adapt to an environment as harsh as the deep sea, humans require many and varied technologies. Scuba diving is not a cheap hobby. When I started technical diving, I thought that there was a point where I would own every piece of gear possible and that I would cease to pour money into equipment. I was wrong—there is no end to this madness. The more experience I gained, the better equipment I needed. When I thought that I owned everything that I needed, then it was time to upgrade my gear. For example, I now buy a dry suit every two years (I alternate between a cold water and warm water dry suit). Because of the type of dives I perform in cold water, I simply do not take the chance of getting a leak in my suit due to wear and tear. A diver can expect to pay $10,000 USD for a new rebreather (although a used one can probably be found for $3,500) and about $7,000 USD for a Diver Propulsion Vehicle (DVP or scooter). With regulators, bailout tanks, dive lights, suits, and other accessories, a CCR tech diver will have between $15,000 and $30,000 USD worth of gear on his back. A second CCR for backup can add another $10,000, and I will not even discuss the cost of an underwater camera system (that could easily cost more than $25,000). Even if divers buy used gear, they spend significant amounts of money to enter into the sport, which means that tech divers generally must have a good income.

For some reason, perhaps because of the cost of equipment, I find that many divers cut corners on the little things, such as not changing their oxygen-analyzing sensors or extending their CO_2 absorbent (sorb) in their rebreather. The high cost of technical diving can therefore be a deterrent and cause the diver to neglect his equipment. The chemical reaction within oxygen sensors analysing the oxygen level consumed by the rebreather diver, and essential to his safety, will diminish over time and should to be replace yearly. The current produced by this sensor may now be limited and unable to provide a high enough reading. Generally, a diver will use a PPO_2 set point between 1.2 and 1.3 at depth. If the sensor milivoltage cannot climb over its limit represented by 1.3, the display on the computer will not go beyond that point. The diver could be breathing a PPO_2 of 3.0 and be at great risk of oxygen toxicity (hyperoxia). Oxygen in great concentrations leads to convulsions and drowning. When divers cut corners on the little things, they are taking risks that could be avoided. In technical diving, a diver must change his mentality from "if it ain't broken I ain't fixing it" to preventive maintenance.

While diving with me on the Jodrey in November 2018, Andrew, my dive buddy, was using a pair of heated gloves with an internal battery pack. Although he was somewhat aware of the risk this posed, the cost of purchasing a better system with an outside battery pack was quite high. During the at least 1.5 hours of mandatory decompression, his glove had a short circuit and started burning. Because the battery was internal to the suit he was unable to disconnect it. The burn lasted over an hour and was so painful that he ended up flooding his suit in 45-degree water to cool his arm. Now cold, he exited the water early, skipping some of the decompression. He suffered a second-degree burn and required surgery to remove the black dead skin which was about two cm deep. However, Andrew did not complain. When I asked him how he was while drinking a pint after the dive, his words were "jme sens en vie en criss"—"I feel fucking alive." He did not complain or whine about his injury, nor was he ashamed of his mistake. The next day he researched proper heating vest systems and gloves with an external battery pack that could be disconnected at any time during a dive.

A Difficult Business

The scuba diving industry is very difficult to make a living in, perhaps because hobbies are often sacrificed first when money is tight. I have observed in the last 20 years a gradual increase in marketing of diving, and especially technical diving. Sidemount is a good example. Sidemount is a configuration that allows the mounting of tanks on divers' sides of the instead of the back in order to push through smaller spaces. In order to do that, the diver needs to sacrifice the manifold that usually links a set of twin tanks, adding the redundancy required to manage the risk of breaking/losing a regulator. When there was no commercially available harness for sidemount, it used to be that only experienced cave divers would make the choice to adopt this configuration, when required by the environment. But in recent years, a publicity and marketing campaign has swept the diving world, making this configuration fashionable. Now there are many dive shops pushing sidemount rigs to make extra money on courses and equipment, knowing that this equipment is not best suited for the environment that most divers are going to risk their lives in.

A backmount configuration is not so profitable for dive shops, since a diver will buy a stainless steel backplate once and keep it for the duration of his career. By contrast, I have seen divers change their sidemount harness two or three times in a five-year span just to keep up with the latest fashion design worn by dive shop owners. There are certain advantages of diving sidemount. It could be that is it cheaper for a recreational diver to start doing technical diving in sidemount as he would be required to buy only a second aluminum tank (if he owned one) and a sidemount harness to get in the game—even if this is extremely restrictive. Thus, it can be financially appealing to the diver, and there are many other advantages to this tool. However, this configuration is not optimized for technical divers in cold water or the open ocean, with thick dry gloves. I see many divers on charters having to abort dives when they realize they are having problems jumping into the current with 3 or 4 cylinders dangling around them. Ultimately, they are also limiting the size of the cylinder, which makes them carry less gas and thus be less safe (though having more gas could allow them to undertake bigger and more dangerous dives). There is a time and place for sidemount, but this also requires practice. I dive sidemount and I teach on the kiss sidekick rebreather. I use this toolset when I need it and when I need to practice, but I will not sacrifice safety for fashion.

The point I would like to make is that when fashion is that important to people practising an extreme sport such as cave diving, it means that those people are trying to optimize how they look rather than to optimize safety. If the focus on performance is removed from an extreme sport, then it becomes a leisure activity, not an extreme sport. In a way, the extreme commercialization of sidemount and the need for beauty is turning the extreme sport into a regular activity. The industry preys on new divers to get as much money out of them as quickly as possible, since "a typical diver stays in the sport only 5 years on average" (Anonymous). By selling inadequate gear to people for profit, some dive shop owners whose judgment new divers trust are complicit in increasing a diver's risk. Once divers realize this, the trust is usually broken with the owners; this is why there are few technical divers in the region that hang around dive shops.

Figure 5 - Diver sidemounting in Mexico (Photo by Steve Duplessis)

Figure 6 - Diver sidemounting in Mexico (Photo by Caroline Joanis)

Diving is a difficult business, especially in a country where the sport is not very popular due to cold water. Only a few at the top of the dive industry will ever make enough money to sustain a decent life. For example, it is now customary for the dive shops in Quebec not to pay their open water instructors, instead giving them discounts on equipment. Most shops would also be secondary work for owners themselves in Quebec, since our cold winters make it difficult for divers to get in the water. So in order to make money in diving, an individual needs to work to invest his or her own resources to get to the top with the help of another income or existing wealth. Attempting to make a living through the passion for diving is a catch-22: if a diver quits his job, he will not be able to afford the diving he wants to do; if he stays in his job, he will not have enough time to dedicate to his passion. Many have tried to become full-time instructors, and some succeed in balancing their lives, but I have heard many stories of burn out and quitting due to the fact that those instructors were simply not doing the diving they enjoyed but instead were showing basic skills to other people.

It is estimated that PADI was worth 1 Billion USD in 2016 (Porter et al., 2016). Why does PADI make millions of dollars in profit every year on the hard work of the scuba instructors who currently are not getting paid? I estimate that it cost around $10,000 USD to become a recreational scuba instructor (note that this varies by country and equipment style). Where is the return on investment in teaching recreational scuba diving, when there is no pay? How can instructors put food on the table? In technical diving in Quebec, the market is simply not big enough to make it full-time employment, although the liability of teaching could put someone into bankruptcy extremely quickly. Technical diving instructors therefore teach the sport out of passion.

It is because of this variance in the technical diver's intent with regard to the practice of the sport that makes it difficult to form dive team; some are there for the passion of the sport while others are there to make a personal profit. Even if the dive shop owner is passionate about diving, the mere fact that he is making a profit on his dive buddies changes the relationship from "equal friend – equal friend" to "profiteer-client". It is this awkward relationship that damages the trust within a team since the question will always remain: Will the profiteer risk his life to help me in a difficult situation or will he leaves me in trouble? Is he only diving with me so that he can make a profit? Without trust, the team cohesion is broken. It is this capitalist mentality, money, and profit that makes it impossible to form a team which is necessary to becoming a deep diver.

Trust and Tech Diving

An important aspect of technical diving is the team one dives with and the community at large. Although some dives are conducted alone, diving is not an individual sport. Whether it is diving on shipwrecks or in caves, a team is often necessary for safety, logistics, and social reasons. A diver learns much more quickly if a more experienced diver agrees to mentor him. When you get stuck in a shipwreck, who will pull you out? When you run out of air, who will bail you out? Who will you share your life-threatening adventures and successes with? The team is an essential part of braving the environment and marks a clear distinction between recreational diving and technical diving. Although competence and good judgment are usually essential qualities for members of a team, those are not the underlying foundation. A team is built on trust. Trust is developed through many challenging dives, sometimes good and sometimes bad. A diver only really starts to know someone when they are facing a situation together where their lives are on the line.

In November a few years ago, a group of tech divers from Quebec consisting of John, Mike, Jean, Roger and myself decided to drive to Virginia to dive a few shipwrecks that were sunk during World War II by German U-boats. Although we chose to go in the offseason, we found the 68-degree water to be quite comfortable and filled with exotic marine life. In fact, those shipwrecks form artificial reefs that now provide a habitat for many species. One of the greatest risks in Virginia is the remoteness of the wrecks; most of the dive sites were a two-hour boat ride from shore, which could be problematic if a medical evacuation were necessary. They are also in the Atlantic Ocean, which is known for bad weather and high winds. The problem with the wind, of course, is the high waves that come with it.

Figure 7 - Divers exploring a wreck (Photo by Kevin Brown)

We were having a string of bad weather, and after a few days of staying dry we chose to try the ocean, since after all we were all tech divers and in decent physical fitness. As a rule of thumb, the captain would not take his boat out when the wind hit more than 15 knots. I asked him if the problem was with the boat and the risk of capsizing, but he replied, "don't worry about the boat, she is made to take the waves. If you can take it, we can go." The problem was really with sea sickness, the operations around the boats, and the risk of a man overboard in a storm.

So we decided to attempt to dive that day, and before even reaching the shipwreck Mike and John were both seasick. We had waves around 6–8 ft, which made it difficult to grapple on the wreck. The problem with waves of that size is that they block the vision from the boat. In a man overboard situation, it is easy to lose sight of the person quickly since with distance, a series of waves appears almost like a permanent wall. The captain had no problem maneuvering his boat, and although he was around 60 years old he could walk over the thin ledge on the side of the boat to set up the grapples even with the boat rocking side to side from the waves. We planned the dive and took extra precautions that day because of the inherent risk related to the weather. My intention was to hook and secure the grapple on the wreck first, since I was the most experienced diver. I would be followed by Mike, and then Jean and Roger would stay on the line until they receive the go-ahead signal. In the event that the diving conditions were too bad, we would "turn the dive." When I got to the bottom, the visibility was about 2 ft—we could barely see anything since the sand was mixed with the water creating a "silting" effect. The ocean was rocking us back and forth by about 6 ft merely from the water movement, and we could hear the ship's steel grinding against itself. The grappling hook was anchored in a thin rusted metal sheet. The pressure of the boat pulling on the line was such that the grappling hook was eating through the metal sheet; there was no way to secure it in place. It was at this moment that I decided to turn the dive, giving the thumbs up to Mike, Roger, and Jean. Roger and Jean understood and started their ascent on the line. Mike was unresponsive to the signal I gave. There are many signals that two partners can communicate with underwater. One of the most important is the "thumbs up," which means that there is a problem and the dive needs to be aborted immediately. There is no room for argument: a thumbs up means that both divers return to the surface since there might be a real threat.

After re-signalling the thumbs up to Mike three times, he shook his head left and right, giving me back a clear "no." After a few moments of him arguing back, what I feared most actually happened: the grappling hook broke free, splitting the thin piece of metal. I held onto it with both hands while drifting, knowing that if I would be blown in the ocean there was a high risk of being lost at sea forever. Mike decided to attempt a free ascent to the surface. When I reached the boat a few minutes later, I signalled to the others that Mike was not with me, so they started to search for him. The waves were creating a permanent wall at about 100 m from the boat and limiting our sight. Then John spotted the tip of Mike's orange inflatable surface marker between two waves. We would lose sight of him hidden by the waves and then see his marker once again.

We got lucky. When we pulled Mike out of the water, his face was completely white. We all learned a few things as a team that day. Mike's attitude was incompatible with the mentality of technical diving. By not responding properly to the turn the dive signals, he not only put his own life in danger, but my life as well. As tech divers, we accept many risks that are unmanageable and we negotiate the edge. When a team member creates additional risk by not respecting a partner's need to terminate a dive, it breaks the trust within the team. At the time, we thought that it was Mike's lack of tech-diving experience that made him behave the way he did, so we chose to give him another chance within the team. But we discovered during the following years that it was actually an attitude problem. As a recreational diving instructor, Mike was used to initiating people into the sport. He could not accept that more experienced members of the team made the final call—his ego got in the way.

In diving, anybody can give the order to abort the dive at any time. If a tech diver does not respect this norm, it can endanger the whole team and ultimately dissolve the trust between people. There is no place for argumentation with this symbol. After this event, we reviewed some of our diving practices and safety equipment. I now carry a GPS beacon, a mirror, and a pen flare in my pocket all the time when doing ocean diving. Some tech divers may believe that this is overkill, but the recent death of a tech-diver friend could perhaps have been avoided if he had carried a GPS beacon to locate him as he drifted off Cozumel.

DIVERS becoming

Beyond the school of diving, where tech divers learn the fundamentals of surviving under water, the process of enskilment takes place both on the physical and mental level. A tech diver's adaptation to his environment can be observed through his movement, his equipment, and his configuration as well as his decision-making. I can often observe and evaluate the expertise of a diver based on physical ability, for example in a demonstration of buoyancy and propulsion techniques. Divers must stay still, hovering in the water column to avoid damaging the environment, their equipment, or themselves. Divers not paying attention to their surroundings and finding themselves up and down uncontrollably may crush coral or stalactite formations that took thousands of years to develop, or they may shoot to the surface and suffer from a decompression accident. As this is a basic requirement to demonstrate minimum control and complete the most basic tech diving courses, a novice diver will often need to correct his buoyancy with small adjustment fin kicks.

Task loading occurs when a series of events arise, either planned or impromptu, and saturate the diver's capacity to manage the chore. The more a novice diver is tasked loaded, the more fin kicks are seen. This is often tested by an instructor during training and by other team members during practice or during incidents. When a diver is oversaturated with critical tasks, he either loses his buoyancy or becomes a victim of the sea. With time, divers should learn to better

41

assess and prioritize their tasks. As a rule of thumb, they ask themselves whether they are breathing and whether it is the right gas mixture to sustain life at their depth. If so, then they still have time to work through the series of events, solving the puzzle one piece at the time.

One example of a bad scenario that divers may need to be able to solve is sharing air through a regulator with an out-of-air teammate while escaping a cold-water mine and becoming entangled in a line. As a diver progresses in skill, he develops muscle memory through repetition until certain actions become second nature. The air sharing will be a reflex, most often triggered by the donor reading the eyes of the receiver even before the out-of-air sign is given. After feeling the line problem and trying to free himself, the diver will know how to drop some equipment like an extra cylinder or to find his knife quickly without needing to look for it. An experienced diver can complete all those tasks without relying on vision, which can become impaired by the elements, and learn to listen and feel the environment, such as the distinct vibration sound and tension of the exit line sliding in a latex mitt.

As a diver progresses in his journey, he learns to negotiate greater depth while staying immersed longer, hence requiring more equipment. In the cold water of the Outaouais region, the need to preserve heat becomes one of the most essential aspects of the dive; by adding layers or warm clothing for insolation, a diver needs to increase the equivalent dead weight to ensure that he will sink. The fixation of the equipment to the diver will need to be more relaxed, giving a less slick look to the diver—a sacrifice not to be taken lightly since cleanliness and streamlining ensures that a diver does not get tangled in lines and swims more efficiently in the current. Nonetheless, the extra thickness of the suit and the diminished range of motion will dictate that requirement, since the diver will choose the security of being able to quickly reach his safety equipment. The novice may not understand why experienced tech divers do not look so clean underwater, but a series of cold-water misadventures will teach them the need to adapt to the environment.

Symbiosis with a rebreather requires time and effort to develop. As the diver moves up and down the water column, his breathing loop volume requires constant volume adjustment for the counter lung to provide adequate breathing for the lung. The oxygen and diluent will in turn need to be balanced to insure a proper PPO_2. The experienced rebreather diver will listen to the noise of the solenoid clicking underwater to confirm the proper functioning of the system. While the novice rebreather diver spikes his oxygen level and utilizes much diluent to flush it down, the experienced diver anticipates the situation and his symbiosis with the machine makes him feel how much he mixes various gases to obtain a perfect ratio.

Laying line is already a demonstration of experience as a diver must master all of the basic skills before attempting to penetrate an overhead environment, where the threats to a diver's life increase drastically. The mastery is easily noticeable in caves: a novice cave diver is often seen fighting the flow of caves where an advanced diver glides naturally, using the flow as an ally. The experienced diver will have an ability to read the stone formation and understand the current, a skill that will be essential to laying lines efficiently in a non-hazardous way to achieve an easier exit. Perhaps one of the greatest attractions of cave diving is navigating in an extremely dangerous environment while following a simplistic line to the exit. As Anna Tsing (2015) put it in *The Mushroom at the End of the World*, we live in a world without the vision of stability, guarantees, or handrails. The guideline laid in a cave shows a simple route to the diver that provides a feeling of safety and control over a terrifying environment.

Figure 8 - Guideline in overhead environment (Photo by Kevin Brown)

Experience and wisdom can also be observed in the mental ability and state of mind of the diver. To perform a deep trimix dive in a cold-water overhead environment, divers need to take great risks. A critical evaluation of these risks is essential for survival. Experience gained by the diver will permit him to estimate how much decompression feels right for a certain type of dive and properly chose what gradient factor to utilize on a computer. He will also have a feel for the severity of the situation when in presence of bends. Are they the result of an isolated incident caused by a too-tight wrist seal, for example, or is the situation critical to the point of require a hospitalization and time in a decompression chamber? This question cannot be taken lightly as there is often only one chamber in a given region, and if one person is using it, a diver with a more serious problem will be left without the possibility of access. This is why it is frowned upon to use a chamber unless significant signs of decompression sickness are present. Most tech divers will at some point experience or witness at least one sign of a decompression incident as a matter of course given the statistics within the sport. All the Outaouais tech divers discussed in this study have experienced or witnessed the bends. In a way, the process of becoming a tech diver requires going through pain and discomfort, whether it comes from the bends, a frozen fingertip, or the weight of the equipment encumbering the body. This is a necessary demonstration of resilience that is an essential part of joining a team.

To manage some of the perceived risks related to performing great dives, tech divers form small diving teams, often with names and branded apparel like t-shirts (e.g., PTO Exploration). The foundation of a team is not the skill level of the divers but the trust that they place in each other. This trust is essential, since a diver needs to know that in a stressful situation, a teammate will remain calm, resist fear, ignore pain, and provide the necessary support to any team member. This is why during the school of diving, a student will often become task loaded with drills to the point where he will accidentally swallow water. One experienced trimix rebreather diver in Outaouais (Andrew) has said, "tu peux pas apprendre sans avaler de l'eau" (Andrew – "You cannot learn without swallowing water", my translation)." An individual learns a great deal about himself when facing a near drowning situation, or a near miss. Through a series of dives and misadventures, a team's trust and ability will develop to the point where they can anticipate each other's decisions. This is why the diving does not end after exiting the water; decompression continues long after the divers are dry and often transitions into a social event requiring food and drink. Tech diving extends well beyond the immersion phase: "la plongée, c'est 50% en dehors de l'eau" (Kyle – "diving—it's 50% outside of water", my translation). This is because the preparation, the planning, the anticipation, and the social decompression is all done outside of the water, and those activities all reinforce camaraderie amongst the team and are essential to building the trust that forms the foundation of the team.

To further understand how tech divers make deep dives beyond the 400-ft mark, I needed to understand how they took and manage the risk of transgressing accepted industry norms, such as breaking the CNS clock, sometimes by five or six times its acceptable limit. The process of becoming a diver has much to do with the balance of risk. In the next section, I discuss how I have adjusted deep diving to cold weather.

Chapter 2. Diving Under

A pivotal moment in my life as a tech diver was the deep dive under ice at Lake Bon Echo. It was the first of my dives that attracted media attention. In March 2018, my team and I broke the world record for deepest dive under ice, which was previously held by Mario Cyr. This exploration dive was of great importance to me as it was a demonstration of the fruits of years of training and innovation in our own environment—this was a Canadian cold-water record. One of the problems I see is that many Canadians travel south to Florida for training in technical diving, as it is perceived as the center of excellence. It is true that there are many competent instructors there, but what they often lack is experience with our northern environment and cold water.

The idea that tech diving in Outaouais is mostly a summer sport comes with current guidelines and rules developed by warm water instructors. It does not include adjustments in the weather such as that one can wear gloves to do cave diving, which is seen as completely absurd in Florida. Adjusting to our environment and sharing that knowledge is what will make a distinct sport in our region. This is why the media attention was critical to this dive in order to convince other divers that tech diving all year round is very possible even in our harsh climate. We thus need to 'bend' the rules, or make some which also adjust to contexts.

Before the Dive

The amount of planning that went into this event was significant. I spent many hours researching the site and its histories. There are a few reasons why we chose this particular lake for exploration. The first and most important was the depth, but the presence of ancient First Nations pictographs was also intriguing. Because of the pictographs, we thought that there had once been a great human presence in the area and that there could be traces of civilization at the bottom of the lake. Because of the depth and year-round cold water in this site, not many divers could attempt this kind of deep dive, so the bottom of the lake was unexplored. This dive was also a test of skills and a validation exercise to see how a team would come together to achieve a common goal. Although this was a non-trivial dive, it was certainly not the most difficult dive that I ever performed. The dive itself had some specific risks, including the overhead ice, the cold water, the lack of visibility, and the depth itself.

I found that the greatest challenge before the dive was the logistics and dealing with some of the personalities in the large support team. During winter, all the equipment tends to freeze and O-rings tear. My regulators frequently leaked, and I ended up breaking a lot of gear. Getting all the team members to work well together was also a challenge.

Fear of Death

Tech diving is responsible for many deaths, especially on deep dives with overhead since many things can go wrong. I have seen my fair share of death and lost some friends to the sport. Some deaths are attributable to equipment malfunction, weather, operator error or medical issues. The tech divers in Outaouais had also had a taste of this bitter situation, and I could feel the anxiety related to the support of a deep dive.

Many divers' deaths have been blamed on the extreme environment, but some of them have been related to medical problems such as heart attacks. It is true that a diver who has a medical issue underwater has a much lower chance of survival because the problem will usually also lead to drowning. For example, during a trip to Tobermory, Ontario, one of the local divers who was sharing our boat had a heart attack and drowned. The same emergency could have happened while he was jogging and he could have passed away on the road somewhere, but since he was underwater he had little chance of resuscitation.

We were all on the ascent line performing our decompression stop after diving the Forest City shipwreck when I suddenly saw a dive cylinder with gear flying down passing two metres beside my head and hitting rock bottom. My first instinct was to think that Pete had accidentally dropped his gear overboard after removing it. So I went down and recovered the equipment, finished my short decompression, and exited the water to find Roger and John performing CPR compressions on Pete. The guys were all exhausted from pumping Pete for over twenty minutes before the Coast Guard's zodiac made it to the site. The Coast Guard crew was unprepared and did not know how to use their own defibrillator. To make matters worse, Pete was a good friend of our captain, who seemed to be in shock and unable to assist with the rescue.

After the event, Roger told me that he saw Pete while doing his decompression: he flipped upside down with his legs completely inflated, just like a new diver learning how to handle a dry suit, and a few seconds later he let go of the ascent line and shot up to the surface. The official cause of death was drowning, although we later learned that he also suffered a heart attack. What we do not know is whether this heart attack could have been avoided—was it caused by too much stress underwater due to something like equipment malfunction, by decompression, or simply by genetics? Pete was already in his early sixties and no longer in his 20's shape.

A few lessons can be learned from this accident: first, listen to the safety briefing of the captain to know where all the equipment is on the boat. Second, carry a defibrillator and know how to use it. Third, carry a pair of cutters to cut open a dry suit. At the end of the day, if a diver suffers a heart attack and floats to the surface with two hours of decompression remaining, the compounding effect of the heart failure, the drowning and the omitted decompression makes his chances of survival very slim. After this traumatizing experience, members of the dive team now understood that death was a very real possibility and that it did not happen only to others.

Planning Meeting

Back to the ice diving in Mazinaw lake - the pre-event meeting a few days before this big dive went horribly. Most of the team and the logistic support helpers attended the meeting, and the plan was relatively simple. The problem was in the attitude of Matt, who was a director in the government. As a novice tech diver, Matt was invited as an observer so that he could learn about technical diving, trimix, and deep diving. He wanted the dive to go according to his plan even though it was well beyond his level of training and experience: he had never done a single helium dive and had no experience in caves or wreck penetration. He wanted me to tie myself to a tethered line just like he had learned in the recreational ice diving course. I told him this was a bad idea for many reasons (typically learned in a diving penetration course), but he would not let it go. As an influential figure among recreational divers, his critique of my plan spread fear and uncertainty in some of the volunteers that were less experienced. At the end of the day, I told him that I was the one taking on the risk of the deep dive and that it was my decision how to proceed.

The Day of the Dive

I woke up at 5am and began my morning routine, which included coffee. There are only a few reasons why I would wake up this early, but respecting my morning routine was very important before such a dive. I kissed my wife while she was still sleeping and left the house at 6:15am to meet Roger. Roger and I had been diving together for a long time now. This, combine with a shared past military experience created a strong trust bond between us. After we finished packing all the equipment for the dive, we departed for Lake Bon Echo at 6:45. During the drive I felt calm and comfortable. To be honest, the most stressful event of all was the CBC live interview the day before. We arrived at the frozen lake around 9:15. Most of the volunteers and safety divers were already there and were checking the ice thickness, discovering that it was thick enough to be able to drive the trucks directly to the dive site and save much time. At the pre-event meeting we had decided we would not take the risk of bringing the trucks onto the ice because when I had done the reconnaissance a few weeks ago the ice was not thick enough. Now, since the ice was more than twice the thickness necessary to support a truck, we changed the plan and decided to drive the truck onto the ice to avoid carrying all the equipment for a kilometre. I took out my GPS and walked over the deepest spot near the pictograph wall, made a small hole, and dropped a line to determine that the water's depth was over 400 ft, as expected. So far, the day was going extremely well.

Matt and two other volunteers arrived late. As they walked towards us, one of them screamed: "NO TRUCK ON THE ICE!" His face was red, and he looked extremely angry about something. I later learned that during the entire three-hour drive, Matt had talked about how my plan was flawed and claimed that he should be in charge of the dive, even though he did not have much technical diving experience. Maybe the reason why people were on edge was that they felt stressed about the event. Matt had only one job for the day: to bring the camera to record the event along with a GoPro that could reach the required depth. A few weeks before the event, he had confirmed to me that he had ordered the special case that would protect the GoPro against pressure up to 600 ft. On the day of the dive, he came empty handed, claiming that he had forgotten the case at home. I learned afterwards that he lied to the group and had never ordered the case, even though he knew that this was critical to proving that the team had reached a record-breaking depth under ice. He knew that without a video, the record would not be validated. Fortunately, I had an older GoPro model that was rated for half the necessary depth but was known to be quite resistant. I used it for the dive, and it did not implode. Matt's ego, his lack of trustworthiness, and his need for control over every situation, including ones where he lacked experience, is what finally made us kick him off the team.

We used a chainsaw to cut an 8-ft-wide triangular hole in the ice and pushed the blocks under the ice to make them melt slowly in the cold water.

Figure 9 - Diver cutting a whole in the ice with a chainsaw (Photo by Kevin Brown)

Jean was in charge of drilling a small hole at a distance of 50 ft from the entry point and sinking a glow stick on a string. If I should get disoriented during the dive and become trapped under ice, this would allow me to find the main hole using the glow sticks as an anchor point for my emergency reel. We tried to set up the tent, but because of the strong wind and the exposure on the lake we were unable to keep it on the ground. Instead, we used the trailer and the pickup box to change into our suits. Then, Roger and John got into an argument with Matt about scooter brands. Matt had bought a few scooters in hope of reselling them to make a profit but Roger had ordered a very similar model from a different brand online. Matt told them sarcastically, "thanks for your support." Roger explained to him that he had saved a $1,000 USD by buying a very similar product and Matt's price was unreasonable.

The team was used to bickering and teasing each other; this kind of petty argument happened before every dive, almost like a pre-dive stress release routine. But this time it was different. Matt took it personally, and the argument started to drive a wedge between the two parties. After all, Matt was trying to turn a profit while we were diving for sport—a major difference. We had all been diving together for almost five years, but with him it was strictly business and had never really turned into a friendship. We were the most experienced divers and local ambassador for the sports, but for him we were another way of making money, which he reminded us of many times.

One of the volunteers managed the divers at the surface and recorded every entry and exit. We started the dive at approximately 12:15pm. Thomas and Kyle, two very experienced trimix diver from another dive team called Plongeurs d'épaves techniques du Québec (PETQ), were the first sets of safety divers and were supposed to escort me to 150 ft. The whole team helped me get ready, clipping the stage bottles and my second rebreather on my side. For this dive I used my Megalodon COPIS rebreather with hardwire Shearwater Predator, a manual rebreather modified with an adjustable orifice valve and a pressure-adjusted regulator so that it is not depth limited. I also used a Kiss sidekick rebreather (with redundant oxygen tank), a diluent of 06/70, a bailout AL80 of 10/54, a bailout AL40 of air (for the dry suit), two Light Monkey battery packs of 20Amp/h with a full heated suit from Gear Canada, a Bare 6 mm dry suit, a 10 mm hood from Waterproof, a custom 21-pound backplate with an extra 8 pounds on the belt, a custom-made reel, an external Shearwater Petrel and a few cutting devices. We also set up an AL40 of oxygen on a hanging cable and an AL80 of 40% oxygen was carried by the safety divers. I had an extra pair of rubber pull-over gloves with a wool liner and a second heated liner glove. I also had available to me the gases of two safety divers at around 200–150 ft during the ascent.

For dives under 400 ft, I find that the redundancy of a second rebreather as a bailout brings a great degree of safety. I know this is somewhat controversial, and some divers claim that a second rebreather is more of a "in theory concept" than anything else, but it brings a redundancy to the loop and the scrubber, which is crucial at that depth. The reason I carried it was that if there had been trouble below 400 ft, my decompression schedule would have increased drastically and overrun my bailout plan. What if I had a malfunction or got tangled in the line? By the time I sorted myself out, I no longer would have had two hours of decompression but maybe four or six hours. Would I be able to resist the hypothermia in this 34° F water? Because of the depth and the cold water, every movement takes time and effort at the bottom, therefore I wanted to make sure that I would have enough gas for most scenarios (though covering everything would have been impossible). However, a second rebreather also brings many challenges, including the introduction of a fourth volume of air to manage at every depth. What is the proper procedure to manage two breathing loops? When should I check the integrity of the second rebreather?

Since I was dealing with an extremely hypoxic trimix, during the descent I checked the integrity of the breathing loop at 60 ft of depth, when the diluent gas was no longer hypoxic. I flushed the unit and made sure that all the oxygen sensors gave me a proper reading. I also carried enough open circuit gases to complete a bailout if my primary rebreather malfunctioned, so the second rebreather was really there as a complete redundancy system. My procedure for a dual rebreather for the time being was to bail out on open circuit, breathe from the tank, prepare the second unit by flushing it with diluent, then switch on the second unit and re-establish a proper PPO_2.

After a quick descent, I reached the bottom and started to explore my surroundings. Unfortunately, there was not much to see other than mud and rocks – no noticeable artifacts from previous civilization. After a few minutes, I decided to turn the dive since the 8-ft visibility was not great at all. On the way up, I met Jean at about 150 ft from the surface and signalled him that everything was going according to plan. I could not find Roger underwater; he had flooded his suit and had had to abort his dive. Two other safety divers came into the water for the duration of the 20 ft stop. After the dive, which had lasted only a little over two hours, the guys pulled me out of the water and assisted me in undressing. As usual, my jaw and lips were frozen and I could not say anything. Although I knew we had just established a new world record for diving under ice at 434 ft, it did not feel like an extraordinary dive.

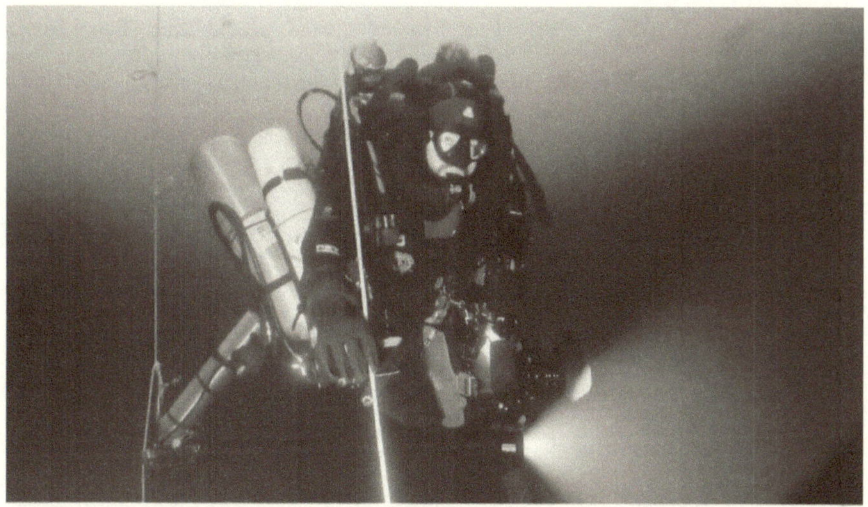

Figure 10 - Myself decompressing under ice (Photo by Anonymous)

Since everyone was hungry and the stress level had decreased with all divers out of the water, the crew quickly worked to tear down the site and leave for the traditional after-dive beer. While pulling the rope out of the hole, Roger accidentally dropped Matt's strobe into the hole, which made things worse between the two. We all made it home that day, and without bends.

A Few Thoughts on Risk

Ultimately, much of this dive was about accepting risk beyond our control. Some agencies train people to dive to 400 ft, like IANTD, but the norm is generally 330 ft. Everything beyond 400 ft is considered experimental, and the scientific research on the decompression schedule for such dives is pretty thin. To add to the complexity, no table has been developed for this depth in cold water. The Canadian military worked on the development of a table going to 270 ft, but nothing deeper. Once a diver has redundancy on every piece of gear, has extra gases, a second rebreathers and many safety divers, executing this kind of dive comes down to accepting the residual risk. It was a unique feeling to realize that I could not control the environment and that my life was at stake: "I knew I was alive." It is a deliberate decision to accept this kind of risk. In a world where people's lives are controlled by many entities, this is what is left for me: to choose to take risks in a borderline environment. When a diver is the first one to ever penetrate a wreck, a cave, or a mine, this choice has always been made. I never know, for example, if the structure will collapse due to disturbances from the gas pressure created by the dry suit dump valve. The uncontrollability of the environment, hence the limit of standards and the need for flexibility, is what I believe makes this sport so extreme.

Why Deep Diving?

The tech diving community requires a reason to perform deep diving. A diver needs to explore a shipwreck, a cave, or anything other than purely going for depth. Going for depth just for the sake of depth is shunned upon and often criticized. I even received hateful comments on social media for making the results of this dive public. It was fascinating to see how quick others were to attempt to control my behavior, making comments like, "you have no reason to go that deep and you shouldn't take the risk." It seemed like many divers felt an inherent right to tell me how to live my life. Perhaps my going to great depth and taking risks that others have decided not to take makes them jealous of the freedom I enjoy? In many sports, athletes push their own limits and can often injure themselves, even dying from pushing too much. Perhaps the number of fatalities related to deep diving was giving it a bad reputation and frightening people. In order to be more profitable, the diving industry could benefit from discouraging divers from deep diving. From my own perspective, an experienced tech diver should be able to perform a deep dive simply because he wants to surpass himself. Going deep just because "you want to" is also an acknowledgement that diving is a sport of its own and not just a method for other disciplines such as naval archaeology, microbiology, or cave exploration. If the community is not ready to accept that, then diving is merely a means to achieve another purpose, and soon divers should be replaced by robots who might be better suited for research. So yes, I deep dive because I enjoy it, for its challenges, its remoteness, its darkness, its total silence and its feeling of freedom. As for people who claim that such dives should not happen because they are simply for the sake of records or fame, rest assured that this is not why I did this dive. Nevertheless, I fully support people who undertake big dives for fame, since this is their own free choice. I understand that many who attempt to break world records, like "Dr. Deep," die trying, and this could furnish arguments for society to regulate the sport—perhaps this is the argument that other divers express when propagating messages of hate toward record divers, or perhaps it is only jealousy. The more the sport is controlled, the more the community will lose as the underwater world will start resembling to the world we live in, governed by strict rules and regulations.

Body, Mind, and Machine: Indissociable Elements of Tech Diving

In 2014, Divers Alert Network identified 146 fatalities in recreational scuba diving (Buzzacott, 2016). At 50 scuba diving deaths within their geographical boundaries, the United States was the country with the highest total followed by the United Kingdom at 12. Despite these figures, new technologies and organizations like PADI have drastically improved the safety of scuba diving to a point where some organizations would say it is a sport for everyone. Therefore, scuba diving does not qualify as edgework, and I believe that it might be more closely aligned with the notion of flow that is also experienced by surfers. Technical diving, however, is a different story. As discussed earlier, tech divers face risks of death or paralysis on every tech dive. If they cannot solve a problem underwater, chances are that they are not coming back. Tech divers voluntarily take the risks and push the limit of their equipment to make longer and deeper dives. Because the threat is present in every dive, I consider technical diving to qualify as edgework.

Lyng (1990) explains that the difference between voluntary risk-taking and edgework is that the latter must include a skill component, which is why activities like bungee jumping do not qualify (although one could make a case that skills are involved in the measurement of the elastic, but this usually does not apply to commercial bungee jumping). The expertise could be an individual's physical or technical capacity, but the single unique skill that is common to all types of edgework is the "ability to maintain control over a situation that verges on complete chaos, a situation most people would regard as entirely uncontrollable" (Lyng, 1990, p. 859). This aptitude is mostly mental and is about keeping attention focused while making decisions quickly, relying on muscle memory and not becoming fearful.

In his ethnography of the test pilot subculture, Wolfe (1979) described pilots' belief that in order to avoid fatal crashes, "the right stuff"—a survival instinct—is required. Lyng found that skydivers also thought that this survival capacity was an innate ability, a capacity to do the right thing without thinking. Indeed, most edgeworkers believe that there is some sort of mental toughness involved in their activity. In the case of technical diving, a great deal of skill is certainly involved. In order to participate in a trimix cave dive, a diver would require several years of practice with months of practical courses and theory. The courses cannot be taken back to back but would need to be spread apart enough for the diver to practice and assimilate the skills learned in each class. For more advanced courses, such as trimix rebreather or full cave, instructors often turn down students who lack the basic skills required. Those basic skills will ensure that the diver can troubleshoot equipment problems underwater in the event of a malfunction. Such an ability is necessary in technical diving, since a ceiling (physical or virtual) is always present, making solving problems underwater essential to survival. A diver who cannot perform a valve drill, for example, would not be able to isolate the equipment malfunction and would risk losing all his gas—a fatal mistake deep in a cave or when two hours of decompression have been accumulated. All of the skills required to survive underwater make technical diving qualify as edgework.

My ethnographic research demonstrated that most technical divers do not agree with the skydivers' perception that someone either has what it takes or not in order to partake in the sport. Most divers would agree that it is the practice of skills over a long period of time that will eventually develop into a reflex, or muscle memory, to respond without thinking in an emergency situation. For example, the S-Drill is the practice of the out-of-air scenario, which is when a diver runs out of air and a fellow diver deploys a second regulator to share air. This drill is practiced every time a team enters the water so that the divers confirm that the hoses are not obstructed and deployment will be in fact possible when in a time crunch. Practicing this drill so frequently also aims at developing the muscle memory of the location of this regulator and the movement required to share it. However, the muscle memory cannot function by itself; it requires a symbiosis of the mental capacity and an in-depth knowledge of the equipment. Mental toughness is a part of technical diving, especially in the cold environment, where the near-frozen water will lock jaws, swell lips, and sting fingertips with pain. The equipment is not comfortable, particularly with the extra weight needed to compensate for the extra thick underwear to stay warm. Most beginner cave divers will lose their finger print from pulling themselves against the current in Ginnie Spring and put tape over to stop the shredding of the skin made thin from long water exposure. Indeed, due to the harsh water conditions and the amount of equipment carried, technical divers need to endure pain even when everything goes right. When confronted with a dangerous situation, such as getting stuck between two rocks in a cave, a diver needs to keep fear and panic in check so that their heartbeat and respiration do not increase, since this would cause them to use more gas in open circuit or increase their CO_2 in closed circuit, which can lead to death. In order to brave such conditions, mental toughness is absolutely required. But mental toughness alone will not make a tech diver survive, it is rather the mix of the body, the mind, and the machine that ensure survival. That's why they are indissociable.

Finally, deep diving is a form of edgework which requires continuous adjustments to the environment. Standards and rules must be bent to allow such adjustments and a strong team bonded by trust will be necessary for a successful dive. Performing dives at extreme depth is not the mean to an end but is an end by itself. While negotiating the edge, a deep diver is more likely to encounter a near miss event, which are essential to the growth of a diver.

Chapter 3. Minex Project – Depths

With a passion for diving the dive team in Outaouais was eager to lay lines in unexplored underwater passages, but the sites that offered long penetration were scarce. This got me thinking. If most of the shipwrecks in the region have been found, what else could we dive? After much research on lost shipwrecks, I started to realize that many of the ships had been carrying minerals, especially in the Timiskaming region and the Ottawa River. In fact, the mining industry in Quebec is extremely important economically, and there were thousands of abandoned mines whose great impact on the land is a concern for many people. Growing up and learning to dive in Morrisson's quarry, I realized that the Outaouais region had a strong history of mining exploitation. In my research, I discovered that the region had many abandoned and flooded mines with great possible depth and largely unchartered tunnels. Since industrial work has not been deemed to have historical value in Outaouais, there is very little documentation of these mines or how the work activities in them took place. I decided to start a project to explore, document, and study those unknown environments, and thus the Minex Project was launched in 2018.

Mining companies exploited sites until they were no longer profitable, then closed down the enterprise. Sometimes, if a mine was below the water line, a bankruptcy meant that the company would stop the water pump and let the mine flood. This is how it came to be that the Outaouais region has a few former mines that now are flooded. The Backmine in Buckingham was a well-known mine frequented by a few divers. It had year-round cold water with a tunnel at 170 ft of depth and a chemical cloud. It was a unique environment and extremely interesting from a technical diving perspective. I had the chance to dive it, but it has been closed by the government;

the entrance was dynamited because it was a popular location for teenagers to party and was a somewhat dangerous place. This was a bad way to deal with the situation: the government destroyed a pristine environment that could have been used not only to develop technical diving but also to conduct research. Also, the presence of divers in any site is often problematic because we never know what they are going to find at depth – keeping a site open just for diving is therefore not a good popular reason.

The Ottawa River also has about three kilometres of known caves with typically poor river visibility and an average depth around 15 ft, but they were already well explored over 30 years ago, and the team was now seeking new challenges. For technical divers, the presence of an overhead in an exploration is fascinating, as it requires lines to be laid so that divers can locate their exit in a maze. What is the difference between being the first to lay lines in a lake and the first to lay lines in a cave? The tech diver's community seems to give more recognition to divers laying lines in caves. In both cases it can be an exploration, but the risk is significantly higher in the case of the cave, making the dives and logistics more complicated. Also, in an overhead environment like a cave, laying lines is mandatory for safety reasons; if there is no line, a diver can be certain that he is the first to explore the passage, and there is a thrill in that.

In the spring of 2018, I called a meeting with the members of PTO (Plongeurs Techniques de l'Outaouais) at the Manchester Pub in Chelsea, where we hold a few of our meetings since as a team we got a discount on our drinks. I explained to the team what the intention for the Minex Project was. Part of the intention behind exploring and documenting abandoned flooded mines was to increase access for other tech divers in order to develop the sport in the region and share our passion. I reiterated that exploration could mean really bad diving or no diving at all, since we had no idea if those mines were collapsed or sealed. At this time I chose not to reveal to anyone the location and name of the first mine—I wanted to know who would commit and join the project first, since some people were new to the team and we had not shared enough dives together to develop strong bonds. Divers are often reluctant to give coordinates of ships or dive sites, since other tech divers could possibly take advantages of a team's research and beat them to the exploration of the site.

I told all the members to let me know if they were interested in the project and gave them a rendezvous point at a nearby McDonald's in a few weeks. After all, the first day would most likely be a reconnaissance of the site, so we would not need anything special beyond all our basic gear. The goal of not revealing all the information up front was simple: I wanted to know who trusted me enough to join in the exploration. I made sure to call the organization who was managing the land around the mines in order to ask permission to dive. Since it was public property, the attendant told me that there were no rules against diving and that anywhere I could take my equipment to the water I would be free to dive. After all, access to water in Quebec is very important.

Moss's Mine

On May 6, 2018, I conducted reconnaissance on the first mine, Moss's Mine, which employed over a hundred men during its apogee in World War I and closed during the Great Depression. The site was filled with the ruins of a small village, and we could clearly observe the vegetation taking over once again after industry had done its damage. I wanted to make sure there was at least water surrounding the site so that the other team members would not waste time coming on a weekend. I found four entry points, all filled with water. When I approached one of the shafts, the unstable ground slipped under my feet and resulted in a fall 10 ft down into the pit.

I gained an eight-inch-long scar on my abdomen, which I still have today, as a reminder that there are great risks in exploration and that it should not be done alone.

On May 12, 2018, four local tech divers showed up at the McDonald's: Ken, Edward, William, and Olivia. After an hour drive north, we had to carry our gear through a punishing walk of 1.5 km to get to the mine's first entry point. We had to make two trips uphill to carry all the equipment that included several tanks and my rebreather, all while fighting starving mosquitoes. Once all the equipment was on site, I carried on with the pre-dive briefing and assigned Edward to dive the first two pits and William the third, while we would keep the fourth one for later as it was quite large. Ken would be supervising and helping William.

While Olivia was an excellent diver, she was not trimix qualified and decided not to dive that day. The importance of her presence is nevertheless noteworthy to mention as it was essential to the exploration itself. In a way, the success of exploration often belongs more to the people supporting the operation than the divers themselves; without them, it would not be possible at all.

I wanted the divers to get in first to give them the excitement of an exploration dive, and I planned to help them as they had supported me during some of my biggest dives. Edward got in first, and after a few minutes we lost sight of his lights; we all understood then that this pit led to an overhead tunnel. I was eager to explore, knowing that there would be some type of penetration in extremely cold water with low visibility. In short, it would be a good challenge. I got suited up and joined Edward, who was now at the surface to brief us. He led the way; by the time we were both at the bottom it was a total silt out. The brown water lent poor visibility, which made it even more difficult to navigate without a reel and a line back to the surface. I followed the tunnel at 50 ft of depth that lead to a bigger room that included many old objects like an old wooden ladder. After going around the room a few times, I discovered a second passage that I chose to follow. I then saw some light. After 400 ft of line laid I exited the water from the mine's second entry point, completing the first traverse. There was another passage, but it seemed like a roofed wooden structure had collapsed and it looked very dangerous to push through. Unlike caves, mines are very recent in history and their support structures filled with nails, metal cabling, and wooden beams are deteriorating quickly. Ultimately, I would need to come back with a sidemount rebreather in order to slip in without disturbing the environment too much—and even then, the pressure of my bubble could possibly be too much for the structure.

Figure 11 - Divers carrying equipment to Moss Mine (Photo by Luc Gilbert)

While I was diving the first entry point, William was exploring the third. He got caught in a web of metal wire, with the spring of his left fin twisted around it as he was breathing down his steel tanks, which means that he was running against the clock to get himself untangled. His Z-knife was completely useless against the steel wire. He had no other choice but to remove his fins and end the exploration. The day was ending and the team was exhausted, our shirts soaked with sweat and our skin full of mosquito's bites. We left some of the bailout tanks on site so we would save time the next day, since I doubted that anyone frequented the place, especially overnight.

On the following day, Ken and Edward decided to dive the largest pit, which was more like a small man-made lake about 200 m long. As they were exploring they made a surprising discovery: an old car from 1970 was sunk at the bottom of the quarry in the middle of the wood. The location of this car was highly suspicious, and this is partly why I think that some people find divers disturbing—we bring back to the surface old things that people have put a lot of energy into making disappear.

Figure 12 - Divers entering Moss Mine (Photo by Luc Gilbert)

After their debriefing, I decided to put on my suit and see if the newly discovered tunnel led somewhere. I tied off my reel to ensure that I would be able to find the entry point in zero-visibility and entered the passage, but after 20 ft I saw the light again. It turned out that this was a U-shaped passageway with both ends leading to the quarry. Breathing on my Meg, I was using a light trimix to be more versatile, so I chose to keep going and explore deeper. The wall was a gradual overhead going to 148 ft of depth. I started to ascend and found a crack that seemed to open up towards something deeper. However, it was too small to explore since I would not fit with my back-mounted rig—I would need to come back with my sidekick, a sidemount rebreather. The rocks also seemed unstable and easily fractured. After much effort, sweat, and pain we concluded the day with a cold beer at the small restaurant near the village. The feeling of being the first to explore a vestige of history that had been frozen in time was amazing. Going down and not knowing exactly where we would end up filled us with excitement. We were all hooked on such exploration.

Forsyth Mine

The second mine site that I chose to explore was the Forsyth Mine, whose proximity to the city of Gatineau made it particularly attractive as a dive destination. The Gatineau Valley Historical Society believes the discovery of iron ore in the region goes back to the first survey of the area in 1801; John MacTaggart, the surveyor, saw the needle of his compass fluctuate in the proximity of what would eventually become a small mining colony. In 1854, the rights were sold to an American firm named Forsyth and Co. of Pittsburgh, Pennsylvania, and the mine was renamed. The ore that was extracted from the biggest iron mining deposit of the region was shipped on the Rideau Canal to Kingston and subsequently to Cleveland, Ohio. The mining operation ceased in 1880, and the activities were eventually taken over by Hull Iron Mines, Ltd. in 1957. It was at this point that the shaft was dewatered and the deeper sections of the mine were dug; the haulage tunnel with rails eventually reached the depth of 200 ft (61 m), and the other deep tunnels were connected via an elevator shaft that was later sunk in 1959. This was the last known exploitation of this mine. Maps I gathered of future development projects for this site were particularly confusing as we had no idea whether they reflected the current state of the mine. The mine and village that used to be 6.2 mi (10 km) northwest of Gatineau City are now abandoned. It sits on the boundary of a recently constructed suburban neighborhood. Nature has claimed back its right, and vegetation occupies the residual foundation of what used to be the miners' houses.

One afternoon, William, Roger, and I visited the site on public land: a small pond with a fence in rough condition that had been cut open. After walking in the surrounding woods for quite some time, we realized that there had been a village here too, as we could still see some of the old foundations. We discovered a second, much larger pond with a rundown fence, but to reach the water we would have required a 30-ft long ladder or climbing equipment, which we did not carry that day. The first small pond did not look promising; there was a concrete slab over the mine shaft opening, but we thought we would give it a try anyway. Instead of getting all the equipment ready, I dressed up in my dry suit and grabbed only a bailout bottle, since it would give me enough air to at least do a first reconnaissance.

I quickly realized that the concrete slab had a crack at the bottom, and I immediately entered; there was a wagon's rail track going down a corridor. A single cylinder would not provide enough safety for further penetration, so I decided to come back to the surface and get the rest of my gear. William decided to join me for the first dive, but since he was in open circuit and limited by the rule of thirds, he would not reach too far into the mine. The water was extremely cold and gave only about 10 ft of visibility. The corridor going down with a 20-degree slope was covered with something green that appeared to be algae, and at a depth of 70 ft I found a thin blue layer that was phosphorescent. I knew that this was an abnormal phenomenon. There was a line installed with a cave diving arrow, marking the presence of a previous cave diver. I followed this line in the tunnel and it led to a small room with heavy machinery, ladders, and an elevator shaft. That was the end of the line. The excitement took over my body, and at this moment I decided to pull my reel, engage the vertical elevator shaft, and lay some line. Upon reaching 150 ft of depth, I decided to turn the dive as I was getting to the limit of the safety of my bailout bottles and I did not want to get into a bad situation.

We got out of the water and started to pack our gear. A member of the group that was supporting the event thought it would be a good idea to bring his pickup onto the trail so that we could load it faster: the great effort of walking with all the gear after long dives increases the likelihood of a decompression accident. The problem was that the pickup was now blocking the path for joggers, and a park manager who saw the situation came to talk to us as he was concerned about the presence of our diving equipment. I quickly reassured him that our dive team was used to these types of dives and that we were not artifact thieves. The fact that we were in plain daylight with our team's t-shirt and logo helped reassure him that we had no bad intentions. I explained to him how we got permission to dive the site by contacting the administration and that the employee had replied that there were no rules against diving (since water is public in Quebec), therefore we could dive anywhere in the park. Ultimately, we apologized for blocking the path with the vehicle, but this incident did raise important questions and we had subsequent conversations with the administration regarding access to water that was part of mines and quarry. There was some confusion in the administration since this was the first time that they were dealing with this kind of scenario, and they seem not to have had regulations in place for something like this. They believed that in order to prevent damage to an environment that could have archaeological and biological potential, they had the obligation to close this site to the public. Ultimately the park asked us to obtain a scientific permit and insurance to continue our diving explorations for liability reasons.

After talking to many archaeologists, I realized that the mine had very little interest for them since its closure was a little too recent and its accessibility made it such that it was not worth the risk. After thinking further on the matter, I realized that the administration's decision and protection of this environment could be because of fear of the unknown. As divers, we bring back to the surface things that have long been forgotten, and due to the proximity of the city and housing, making an unpleasant discovery such as contaminated water could have severe snowball effects. There is also a liability risk that is not fun to share: nobody wants to have a dead diver too deep for the police to retrieve. Overall, it is bad publicity.

A few months after the dive, I got in touch with a microbiologist, Dr. Cassandre Lazar, through a contact at the Explorers Club (a group of people who shares a passion for field science and personal challenges). The first discussion was enough to convince her that we were looking at something unique, and she mentioned that those "algae" in the room were not in fact vegetation since there was no light or heat and little oxygen; these would most likely be bacteria colonies. We both agreed that this mine should be investigated further and that we needed to take samples to analyze in her lab. We still needed a site access permit with the administration in order to continue the research. Upon reaching out to a few researchers, I discovered that one of them was currently performing research in the same area and had a scientific permit, so we started to plan out a collection of samples of the water and bacteria colonies. The park's administration ultimately asked us to apply for a permit under our own name, so we did.

Deep Diving for Microbiology

As the surrounding aquatic environment had been abandoned since the mine closed, microbial communities had thrived and adapted to the conditions found in this underground ecosystem. This constituted a unique opportunity to explore and discover new life forms and rare micro-organic communities in this atypical environment. The dive team was able to find old pictures and rudimentary maps that no longer reflected the state of the Forsyth Mine; documenting the passage was therefore key to the safety of the diving operation to allow the collection of DNA samples. We collaborated with the University of Ottawa and the University of Quebec in Montreal (UQAM). The Explorers Club endorsed this expedition and sent us their official flag (serialized #101) to represent them, with the aim to assess microbiota diversity within the flooded mine ecosystem. Furthermore, the exploration of this underwater mine provided an opportunity to study deep diving in an anthropological way. The microbiologists thought that the data collected could possibly lead to important changes in our understanding of the tree of life (the tree of life is a research tool to describe relationships between organisms based on Darwin's theory); the divers just wanted to dive and lay lines. As an anthropologist collaborating with microbiologists, I had to read on some of their foundation theories to understand their thinking. On his voyage as a naturalist on HMS Beagle, Charles Darwin made a series of observations that would, upon his return to London in 1937, lay the foundation of his work on his theory of evolution. After many years of reflection, he was convinced that species were not created independently; instead there was a lineal descent, and natural selection was the primary but not exclusive explanation behind evolution (Darwin, 1859)[17]. As outlined by neo-Darwinism, the concept of natural selection was a key component of evolution. This "survival of the fittest" would favor individuals with the phenotype that provided an advantage in their environment over the other members of the species. These individuals would have acquired this different phenotype through a random mutation of genes. Because of the success of these individuals and their enhanced ability to survive, they would be more likely to reproduce and transmit that phenotype. Over time, this mutation of genes would no longer be the exception but the norm in the phenotype of the organism[18].

From a diving perspective, what we observed was fascinating; the formations resembling long stalactite algae or white filaments spread in a spider's web were surreal. We knew that things that grow without light, in 42° F (5.5° C) water, deep underground, and with little oxygen would be resilient and possibly unique to this kind of environment. Studying and understanding the resilience of life in an environment that has been exploited by humans and abandoned when profits plummeted was almost like having a small crystal ball that showed a little of the past, presents and futures of planet Earth. Finding new life forms with different metabolism mechanisms in those harsh conditions at depth could be of great importance for the microbiologists in the team.

First Sampling Dive

With a team of five divers, we started collecting samples in winter 2019; this task was particularly challenging due to the year round cold-water element and the large number of samples required to perform a credible analysis. The original research permit was expiring at the end of March 2019, which meant that diving in Outaouais would normally be under ice. Since we were pressed for time, I went to perform reconnaissance at the site to evaluate whether it was possible to dive under those conditions and whether the ice had the appropriate thickness (when it is thicker than the chainsaw blade it becomes very laborious to cut a hole open). We knew winter and extreme cold weather would bring its share of malfunctions and problems, which is why there are almost no technical divers performing difficult dives in such environments. Upon arriving at the dive site I discovered that there was no ice covering the pond connected to the mine. This was

74

quite unusual, since we had over 18 in of ice at Morrisson's quarry, a dive site that was only a few kilometers away. I thought that the water from the depth must be working like a thermopump system preventing the surface water from freezing.

Figure 14 - Forsyth Mine entrance without ice (Photo by Anonymous)

We all meet at Roger's house around nine in the morning on February 18, 2019, all a little anxious about the unknown factors of meeting and collaborating with others. I chose to arrive an hour earlier just so I would have some private time with Roger to go over the diving details since after all I was going to conduct a dive to 300 ft of depth in an overhead environment in freezing cold water. Dr. Lazar and her student, Xavier, showed up carrying the samples bottles in a white pickup truck with the university's name "UQAM" on the side. After we introduced ourselves, I looked at the sampling bottles and realized that they were glass and not very thick. I knew that fragile materials often break during dives—and indeed we did break a few. Nonetheless, we proceeded to secure rock-climbing carabiners to the top of the bottles so they would be easier to carry. An hour later, an anthropologist from University of Ottawa who had a research permit for the area, Dr. David Jaclin arrived with his friend who was going to film the entire process. We proceeded to get dressed in Roger's garage in close proximity to the mine and readied our equipment since the day was cold in the middle of winter. I was answering David's questions on tech diving and equipment while getting dressed; I explained that I frequently changed my gear because I do not want to take the risk of diving with old equipment, especially dry suits (which I change every two years) since they become more prone to leaks. Roger quickly replied: "No that's not true, he changes his suit every two years because he is becoming fat! You know Kev, we need to tell the truth to each other…" We all laughed and the stress and pressure of the dive slowly diminished.

It was only the second time that I was meeting Dr. Jaclin, and although I did not know him well I had heard great things about him and his willingness to help. Roger and I quickly became convinced that he was the right fit for the team when while eating a pastry I dropped a pecan onto the garage floor. Dr. Jaclin subsequently picked it up and ate it, commenting, "this is good, we can't waste it." As an intellectual, he passed the test to be accepted in a group of tech divers that is very closely knit. Later that day Roger told me while laughing that David did not know what his dog had done to the garage floor!

The trucks were already packed and we were dressed, so we quickly drove to the site. We took the toboggans and loaded them with bailout bottles to slide them on the thick snow, shouldered our heavy rebreathers, and started walking in on the trail. With my Weezle Extreme Plus underwear, my heated suit, and my low-pressure fifty tanks mounted I was heavily laden. My gear weighed over 100 lbs, and as a result I was breaking through the snow and sinking to mid-thigh. Unable to walk, I lay on my back until Dr. Jaclin came to help me carry my gear on the toboggan.

We entered the pond and rigged out stage bottles. The temperature at the surface was around 43° F. I was carrying my Sony AS7II camera, five sampling bottles, and three cylinders of bailouts—I could hear the bottles tinkling together while I was swimming down the tunnel. I dropped the camera in front of the elevator room so I could go down the shaft. I made my final check around 90 ft of depth and engaged the descent. The problem was not the descent but rather the ascent. I was not really bothered by decompression or cold water since I was used to those factors and prepared for them. What had kept me up at night was the thought of venting my rebreather coming back up, creating massive bubbles pressuring the ceiling of the mine and potentially causing it to collapse, making it my tomb. Being stuck behind a pile of stone blocking the only exit is not a happy thought. This is when I tossed the dice for the first time during this exploration. I evaluated all my known risks, understood them, took measures to mitigate some and accepted others. I knew there was significant residual risks for this dive. "I won't know unless I go – I'm doing it." I went down to 300 ft and realized that there was another tunnel penetrating deeper into the mine with a number '10' painted on the side of the wall. I collected the sample with a syringe. The shaft was made for two elevators side by side with a third section of wooden ladders, probably so that the miners could escape in case of emergency.

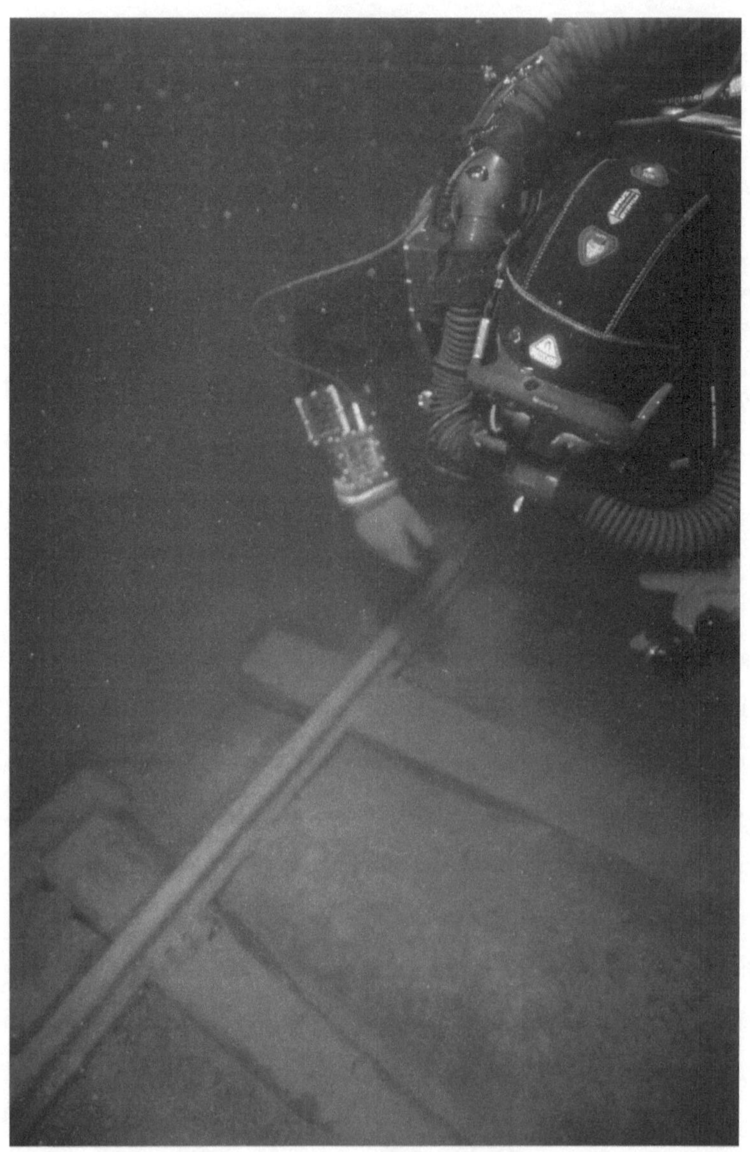

Figure 15 - Diver following the train track for guidance (Photo by Kevin Brown)

Going up was a challenge. While ascending I had to purge my rebreather that was inflating due to the pressure decreasing, and those bubbles created a whirlwind disturbing all the sediments on the walls and resulting in a total loss of visibility. I then saw the stone wall starting to move and slip, and my heartbeat started to increase—I thought that part of the mine was starting to collapse. I tried to move up but the ceiling was stopping me from exiting. In such a situation, adrenalin starts pumping until I accept the fact that I may not get out. Then I focus on my tasks and try to escape. While I was staring at the stone wall, the sediments dropping in the shaft created an optical illusion, similar to when someone sitting in a car at a red light sees another vehicle on the side accelerating and thinks their own car is moving backward while it is in fact stationary. This illusion was extremely disorienting, and an increase in the heartbeat could be potentially dangerous for decompression sickness. I held steady and focussed on my breathing to reflect on the situation. I then realised that I was stuck in one of the wood beams solidifying the walls hitting my head on what I perceive as a rock that had fallen and blocked the only exit, but was in fact the wall. After realizing that there was also a rock now obstructing the ascent, creating a restriction, I was able to exit the illusion shaft. I collect the rest of the samples while finishing decompression, and exit the water. The camera and the five sampling bottles kept getting stuck in every obstacle during the dive. Only Roger knew that something had happened down there, as he can recognize such emotion in my eyes. The others had no idea of the risks we were taking. Xavier and Dr. Lazar proceeded to filter the sample and left for Montreal so they could freeze them in order to preserve the DNA. The rest of us went for the traditional beer at the Chelsea pub.

It took me a while to understand what had happened during that dive. At first I thought it might have been caused by an accumulation of CO_2 in my rebreather, but it was not. The following weekend I returned to the mine with Jean, but this time I reached a depth of 453 ft, just for the purposes of exploration and mapping: I discovered three other tunnels and lined the main shaft in preparation for the big dive.

Figure 16 - Divers descending in the elevator shaft with debris (Photo by Kevin Brown)

The Big Dive – A Descent to 602 Feet

The idea and execution of a deep dive is not improvised, it is planned ahead. It took me only one dive to realize that the elevator shaft descended to great depth; I knew I had to go see for myself. I established my goal of 600 ft, as it was realistic and achievable while still requiring a full year of preparation both mentally and physically. It required me to push my limits and go outside my comfort zone.

I took the time to write some notes while I was preparing for this challenge so I could compare my state of mind during the preparation and after the accomplishment. I wrote the following reflection on June 21, 2019:

"I am currently at a natural point in my life where my experience is good enough and my skills are still sharp. I also do not have any children, so my responsibility towards others is limited. If I wait to do this dive and I start a family, my diving skills could fade due to lack of time for practice. It is that perfect balance between skill, experience, and youth that makes me believe that this is the time to execute a deep dive. Although I have been weight lifting for many years, four months before the dive I started to increase my running, averaging over two hours per week. I am still not where I wanted to be cardio-wise, but it's decent enough. I could lose another 10 pounds of fat as it increases risks of DCS, but it's also a natural insolation for cold water. Being a little fat makes you warmer. My kit is ready, all gases have been blended and analyzed by myself, there will be no one to blame if this goes wrong. I understand that I am walking into a dive with approximately 33% chance of failure, although this is an educated guess since there is no way to know if this is accurate or not.

I have been practicing for this dive for over a year now even if not many people are aware of it. I can count many sleepless nights thinking about it. In a way, I have been stuck in time since the day I started diving this mine and learning the potential for deep diving that it offers. To move forward, I will have to pass this challenge, to solve this puzzle. It took me time to accept it, that there is a chance that this could be my last dive, but I have prepared for it. I have written my will and final instructions; by doing so, it will help keep my mind focused and remove some anxiety about death.

If this is to succeed, it could be a world record for mine diving in cold water, with no decompression bell (self-reliant). It would be at the minimum, a Canadian record I believe. Records and recognition are important to humans, but it does not bring happiness and it is not the reason why I am doing this dive; I would trade any world records for just a regular dive with close friends and the camaraderie of the traditional beer afterwards. This dive is about a man against himself, a boxing match where the opponent is an exact replica of yourself.

What is life worth if you cannot do what you enjoy with the people you like? I have often tried to justify my deep diving passion, for example, by taking samples for microbiologists to analyze and do scientific research. I see deep diving as a finality by itself, and not a method to achieve results in another discipline. But deep diving was shunned upon by some diving communities, alienating those who choose to go for depth "just for depth." A source of justification could be that it was very difficult for those communities when a friend passed away in attempt to break a record.

When preparing such a dive, you have to question yourself about death and reassess your perception of it. I cannot say if death is the end or the beginning of another journey. I don't really know if I will be self-aware, or just a pile or dirt somewhere. At this time, the question I am asking myself every hour of the day is if that mine will be my tomb."

Before the dive, I attempted to convince myself that if it went badly, my legacy in the sport could live on and in a way could enter the memory of humanity—which is often enhanced by the death of the contributor. In retrospect, I find that those thoughts were organized in a way to ease my acceptance of the extremely high risk related to the dive. Caring about leaving a legacy behind makes no sense to me, as I would not be present to see its influences on the sport. The only time I could enjoy this feeling would be the time between the moment I realize I will be dying and the time of death—if such a moment exists.

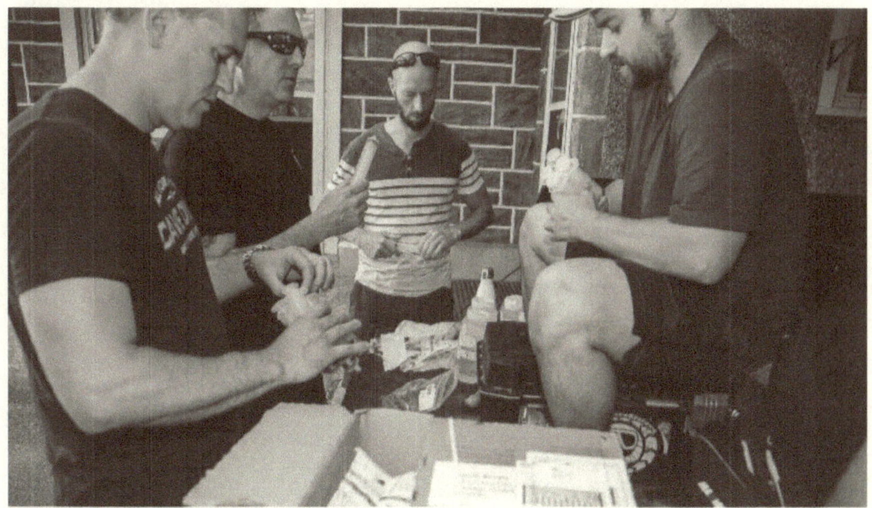

Figure 17 - Divers preparing sample kit before the dive (picture by Anonymous)

Preparation

There is nothing simple about executing a cold-water dive in an overhead environment at a depth of 600 ft. Every dive in the preceding year was aimed at developing and validating the skills and techniques required for this environment. I decided to change my rebreather configuration specifically for this dive, as small cylinders would not provide a sufficient quantity of gas. I also decided not to carry a second rebreather, as it was easy to stage multiple decompression cylinders and there was not a great requirement for linear penetration that would have dictated an enormous amount of bailout gas. However, the great depth and atmospheric pressure that came with it did require a significant amount of diluent gas for the descent and, if needed, to flush the rebreather. For the ascent, I would need much oxygen just to keep the PPO_2 high to ensure efficient decompression. This is why I fitted my rebreather with two 50 ft^3 low-

83

pressure steel cylinders where both regulators would be fitted with a QC6 connector to plug into my bailout valve. Of course, the diluent was plugged into the bailout valve for the entire dive, but I wanted to have the oxygen also available, since I would not want to be stuck underwater with a bunch of decompression time to do, a ton of oxygen in a tank, and no way to access it. I would voluntarily overfill those tanks in a "cave fill" fashion, pumping them to 3500 psi to carry extra gas as opposed to a normal low pressure 2400 psi fill. I was not aware that anyone else used this kind of configuration at this time, but it worked for me.

Another aspect of my preparation was mental. My mental readiness was developed through a full year of visualization, where every night I would imagine the possible scenarios and the movements required to survive. If a diver's mind is not ready, then the dive should not be attempted. I also eliminated almost all alcohol consumption, resulting in good weight loss. Physical and mental fitness are essential to deep diving. Even if science does not understand decompression completely, and surely cannot explain all of it, we do know that physical fitness has a great impact on the quality of decompression.

To maximize my chances of survival, I needed to do everything I could to plan and prepare myself. For this purpose, I used a progressive penetration strategy to slowly assess the solidity of the environment and its risk. I did not rush to the depth of 600 ft, instead starting at 150, then 300, then 450 ft. Not only did I learn how the shaft was built and which obstacles were lying where, I also memorized restrictions that I would have to pass in a zero-visibility exit.

A week before the dive, I generated my decompression schedule based on a theory model. After all, those models are extrapolated from tests on military divers, but none of them were ever done in cold water at such a depth. They are therefore experimental and I am the test for them. I decided to use a 5/95 mix for my bottom mix, which was 5% oxygen content and 95% helium – a deadly hypoxic gas if breathed at the surface. My gear was assembled and tested at least 48 hours in advance; everything that could have been prepared had been prepared.

A few days before, I had met with Richard, a local technical diver that have been diving with me for a few years. Although less experienced in rebreather diving, he was quite competent, with plenty of experiences in other extreme sports. We went through every detail of the plan as he was going to be present that day. He understood the desire to deep dive.

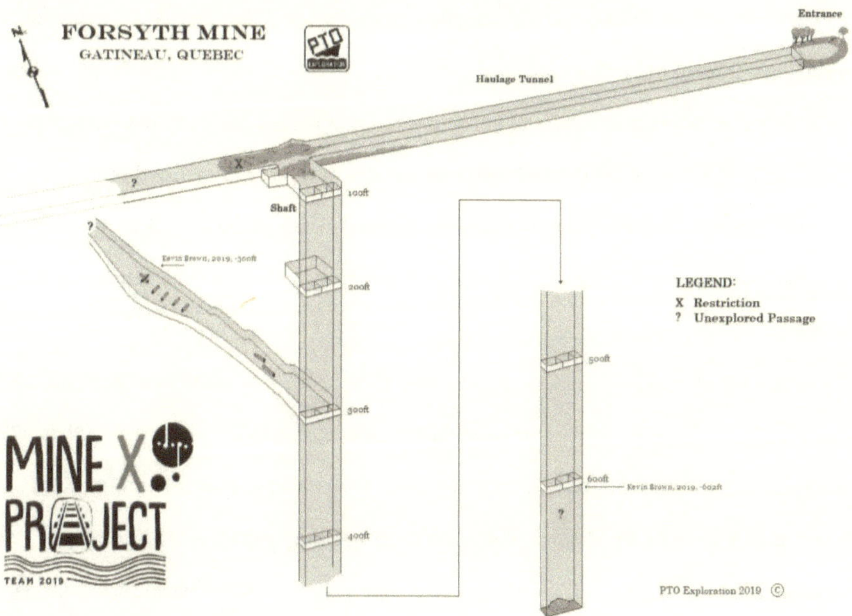

Figure 18 - Map of Forsyth mine

The Morning of the Dive

On the morning of July 20, 2019, the team met at my small apartment in Hull for the day's briefing and final preparations. The heat was painful. It must have been at least 35° C, and we sweated while standing in the shade. Jean and Bob were going to dive with me; we prepared the sampling material and discussed the plan. Although Bob and I only started diving together recently, it was easy to trust him; he was an accomplished diver with over 35 years of scuba experience. All divers in Forsyth's project shared the commonality of serving time in the military. I had initiated Jean to scuba diving while being located at the same military base over ten years ago. Sharing life experiences like going through basic training reinforced the trust between us. Jean would take the 20, 40, 60, and 80 ft samples and Bob would follow me past the restriction to the elevator shaft so he could explore that room until my return; he was hoping to find another tunnel to the lower levels. Richard was going to wait for us at the surface and was ready to react in case of an emergency. After we dropped all the stage cylinders in the pond near Forsyth's entrance, I donned my red Seaskin dry suit with my heated undergarment for which I was carrying and staging a total of three battery packs, two 20 amps and a 30 amp. The guys helped me rig my Megalodon Tiburon rebreather.

After a full year of preparation for this deep dive, I entered the mine, which had no more than 6 ft of visibility, and followed the intact train track as guidance. After only about 400 ft (122 m) of penetration, I passed the first restriction, a small squeeze between an industrial pipe and the floor's mud leading to the elevator room. The main haulage tunnel was clogged by years of accumulated sediment. That old dual elevator cage, where miners used to lift the ore, was the only way to access the deeper levels of the mine, which ran every 100 ft (30.5 m).

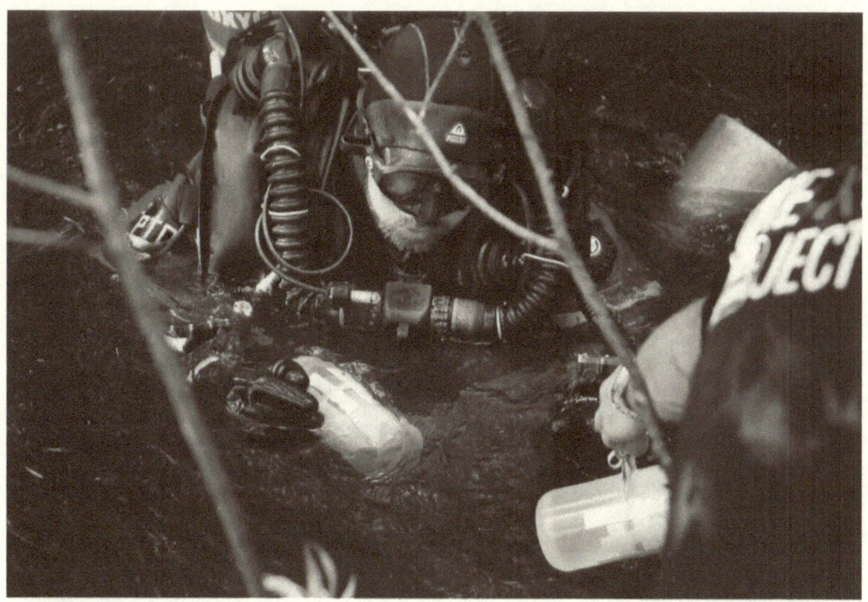

Figure 19 - Diver taking the sample kit before the dive (Photo by Luc Gilbert)

Earlier in the project during my first dive in that shaft I had been anxious; we thought that the mining company had sealed the elevator shaft from the surface with a concrete slab, but I had no way of testing the stability of the potential underwater land fill. I knew that once I would be returning from the deep, the bubbles from my rebreather would accumulate and create pressure under the landfill, possibly causing some parts to collapse and seal shut my only way out. In my first descent, some of the wood planks came loose and fell on my head – which was not that critical at the time. If that were to happen in a dive that accumulated 6 hours of decompression in 42° F (5.5°C) water, a single nail from those planks could rupture my dry suit and put me in a difficult situation having to choose between severe hypothermia or a decompression accident; both could lead to death before exiting the mine. I knew there was no way I could stay in water this cold for over an hour in a leaking suit, even with a heated undergarment. This was the risk that I had to accept to make this dive.

Before the dive, I had tried to estimate the risks. As an engineer, I feel more secure knowing that I have examined the math and statistics of a situation. I estimated, ridiculously, that I had a 33.3% chance of success, 33.3% chance of aborting, and 33.3% chance of failure; failure here was defined by a severe type two decompression illness, permanent paralysis, or death. Such calculations put my dive buddies at ease since making a detailed risk analysis and coming up with a final number for the success, abort, and failure likelihood demonstrated all the thought I had put into this dive. The thing is that there are so many unknowns about this dive regarding the environment that it made the statistics a complete shot in the dark. Nonetheless, the exercise was critical to the success of the dive as my mind could go on to focus on the mission. Placing a number on a risk makes it so much more acceptable as opposed to leaving it unweighted. Also, knowing that my team would be able to say, "he knew the risk," in the event that I had a bad day was important to me, so in a sense this exercise protects their mind as well. For my part, it is a short lifetime of accumulated experiences through a dedication to diving that made me survive a dive like this, preventing situations before they even occurred and managing unknowns as they arrived – although I recognise that chance had a big role in it.

Figure 20 - Diver posing in the shaft (Photo by Kevin Brown)

I checked my gear for the last time and started my descent into the shaft. Everything that I could have planned had been planned, everything that could have been practiced and rehearsed had been. As I passed the 150 ft (46 m) mark, the visibility improved to about 50 ft. I had previously dived it and installed a line up to 450 ft, but descending another 152 ft (46.3 m) would be challenging since the shaft's unstable structure was filled with debris, rocks and metal cables—objects that could rupture my equipment—creating multiple restrictions along the way. Because of this debris, my descent speed was too slow, which resulted in gaining an extra hour of decompression. I needed to analyze and memorize every single detail on the way down because I knew that on the way up, purging my suit and rebreather from such depth would create a whirlpool of silt, making it a zero-visibility exit. I must avoid touching or disturbing anything since the structure could be unstable.

Upon passing 500 ft, I observed that a section of wall in the elevator shaft was caving in; the wooden plank seemed to slope towards the middle of the shaft, and others were clearly no longer attached to the main structure. I knew that I was the first human to ever pass that restriction and I would also be the first one to create a pressure bubble on my way back. Because of the extreme depth I was accumulating significant decompression time, so there was no time for a detailed estimate of the structure and I decided to bypass it and make a leap of faith.

I finally reached just beyond my planned depth, at 602 ft (183.5 m), but the laborious descent took more time than I anticipated due to unforeseen restrictions; I would therefore have only a few minutes at this depth. Nonetheless, I took a few moments to observe the environment. Although it felt like minutes in my mind, I am sure that it lasted seconds. In that borderline milieu hostile to life, time becomes relative to the perception of the individual who visits. Everything was dark, and the wooden structure appeared to be burned from a fire, though perhaps it was a bacteria that produced this effect. I pulled out the syringes to gather the DNA samples. I could feel the high-pressure neurological syndrome (HPNS): my hands were not steady, and my eyes were having difficulty focusing. This was further complicating the laborious sample collection with my dry mitts. This place felt like a dead zone, hostile to all life, where even technology barely allowed humans to visit. Finding microorganisms that survive in such environment of pressure, darkness, and cold would be fascinating; it would be a great demonstration of the resilience of life on earth.

I began my ascent. I could hear the clicking noise of my solenoid indicating that my rebreather was injecting oxygen to compensate for the gradually diminishing atmospheric pressure, resulting in me purging the gas that became poor in oxygen to allow optimal decompression of my tissues. As the pressure diminished in the water, the density of the gas in my counter lungs also faded, making the helium pressure in my body superior to the one in my breathing system. Acting like a filter that attempted to restore the pressure equilibrium between two systems, my lungs were purging the extra helium. I started my first decompression stop around 450 ft of depth, and my computers indicated over six hours of total decompression time. I choose to change the gradient factor on my diving computer to accelerate my ascent and remove some of the added contingency time—a decision that increased my likelihood of getting bent. I only had a small pinch in my left shoulder caused by bubbles stuck in my tissue, a feeling that I have had far too often, but it passed around 150 ft when I slowed the ascent slightly based on gut instinct. My left shoulder was always a problem while decompressing; the scar tissue from my time in the military must make the passage of bubbles more difficult, like an invisible tattoo.

I knew that I had to negotiate the nasty restriction to re-enter the main haulage tunnel and exit the elevator shaft, a place where there were always risks of rupturing my dry suit, especially in a zero-visibility situation. But it went well. In the main tunnel, the stress diminished as I saw the familiar faces of Jean and Bob and the cylinder of extra decompression gas marked 50% oxygen. Because I still had four hours of decompression, I left the extra cylinder in place; in a typical dive I would have picked it up and carried it with me to the exit, but every time I touched a clip there was a risk of puncturing my gloves, which would have been catastrophic at this point in the dive. The last four hours of decompression were painful; because of the extra thick diving underwear and heated suit, I was wearing a 21-lb backplate, a 6-lb stainless steel rack, and 8 lbs of lead. All this weight was hard on my back, fighting the buoyant force of my suit. I finished off my last two hours of decompression at 20 ft of depth, where I could almost see the light from the outside world, but then I lost feeling in my left hand, which was completely numb. At the time I was unsure of the cause of the numbness; it could have been the bends, the cable of my heated glove digging into my wrist, or simply too much time in cold water. There was nothing to do but wait in total silence.

Figure 21 - Elevator shaft (Photo by Kevin Brown)

Decompression profiles are more like a curve: the deeper stops are often very short and the shallower stops longer. At depth, I used the time on decompression to think about what was coming up and prepare for the next phase or ascent. In the shallower stops, such as at 20 ft, the risk of equipment malfunction is reduced as I sit still and only use pure oxygen in the rebreather. The helium off gazing from my body was even cutting down my pure oxygen too much and regular flushes were required to keep a decent partial pressure of O_2 in the breathing system. Nonetheless, the shallow stops of decompression are precious moments where I get to have time with myself and nothing else to do, nowhere else to go. In a society that seems to be always chasing time, decompression seems paradoxal; it is a situation where there is too much time and the task can only be completed by slowing down.

I exited the water after 6 hours and 13 minutes. I completed another twenty minutes post-breath on pure oxygen relaxing in the warmth of the sun in the sump water of the small pond while listening to the sound of frogs. A nasty smell, similar to rotten eggs, woke my sense of smell, reminding me that it had been completely absent for the last few hours. My knees were sore, my whole body was drained of energy, and Jean and the two microbiologists had to help me walk out of the water. But I did not bend, and the dive was a success. I handed over the water and sediment samples to Xavier, who would subsequently analyze them.

There was a distinct feeling of accomplishment after dive like this. I do not know how to describe it other than that it was a mix of accomplishment, happiness, endorphins, exhaustion, and calmness.

Figure 22 - Myself after the dive at 602 feet (Photo by Anonymous)

Exploration Results for Forsyth

In all, we have explored the Forsyth Mine elevator shaft to 602 ft (183.5 m) along with the 200-ft level and the 300-ft level. When the visibility was at its best, we could see that the 200-ft level consisted of a large room approximately 50 ft (15.2 m) by 30 ft (9.1 m) with a few industrial materials lying around. Towards the end of the project I explored the 300-ft level, but because of the difficulty of running a reel with dry mitts and the lack of natural elements to tie the string onto, I only made it to about three quarters of my planned penetration; as the research permit was

94

expiring, I could not return and confirm the end of the tunnel, although I had collected enough remote samples.

Due to lack of time and logistical complications, the 400-ft level, the 500-ft level, and the 600-ft level remain unexplored. There is also one section of the main haulage tunnel past the elevator shaft that was obstructed with silt that we choose not to disturb. We believe that the exploration dive at 602 ft in 42° F water sets a new world record for deep diving in a cold-water mine (autonomous—i.e., without decompression environment)[19]. At this time, no further exploration of the mine is planned.

Figure 23 - Microbiologist testing the water (Photo by Luc Gilbert)

This exploration allowed researchers to collect a total of 33 water samples and 62 biofilms (gelatinous matrices secreted by microorganisms to favor their survival). Genetic material (DNA) present in the samples was then extracted at the UQAM labs, and these are now awaiting sequencing, which will allow researchers to precisely identify the family of microorganisms present in each sample, deepening our understanding of the range of biodiversity present in the unique ecosystem. Preliminary results of DNA concentrations are promising: there is DNA at depth. The microbiologists expect to find well-established non-phototrophic (not using light for energy) microbial communities within the mine. Furthermore, the variety of biofilm forms observed and sampled gives hope for the discovery of unidentified microbial species[20].

Figure 24 - Bacteria colonies on the locker (Photo by Kevin Brown)

Controlling the Uncontrollable

Research on control has been central to psychology for quite some time. One finding is that it is detrimental for individuals to believe that they are in control of their own fate to maintain sound mental health (Abramson et al., 1978)[21]. In technical diving, the illusion of controlling the environment can be fatal. A great deal of skill is required to perform the dives, but in wreck diving for example, the chance that a strong current caused by a change in weather can blow a diver away is always present. There is a point when divers forget or accept that this risk is always there. Divers cannot completely control the outcome of the dive, since there will always be an environmental factor, but they will be able to control the most likely outcome—it is an extreme sport after all.

Batchelor (2007), who studied young Scottish women who were victims of violent crimes, found that risk-taking was a means to regain control in someone's life perceived as out of control. By seeking risk, these women would feel and remember that they are "alive," making risk-taking an essential survival strategy. Similar to the psychological theory, Lyng (1990) found that there is an illusory sense of control in edgework that counters the absence of control in an institutional routine. It could be that the planning and long preparation involved in such activity would contribute to creating an illusion of control. Using Lock's (2019) turkey example[22], we could say that a routine does reinforce the illusion of a forcible outcome, even when uncontrollable. Lyng

(1990) presented a paradox in edgework: people who feel helpless in the face of events beyond their control (socio-ecological) turn to experiences that are life-threatening and more dangerous.

Sensations are an essential part of edgework. Lyng (1990) found that at first participants claimed that the experience produced a sense of self-realization and self-actualization, and that there was to some degree an initial feeling of fear that needed to be confronted and controlled. He mentioned that this contradicts the stereotype that risk-seekers are fearless, but this fear later yields to a feeling of excitement and omnipotence. This feeling of extreme power from the victory over the challenge provides a sense of control over the participants' lives knowing that they can face any threatening situation. Lyng (1990) concluded that it is this feeling of extreme power that contributes to the creation of an elitist culture in edgework. The participant's perception and consciousness is also changed during edgework; the feeling of time passing is distorted, the field of vision is narrowed, and a feeling of oneness with the equipment or environment is created—as when a motorcycle racer would claim to be one with his machine exercising mental control over the bike (Lyng, 1990). This feeling of becoming one with the environment can be explained by the altered senses making a connection with some object, as if the participant could control it mentally. A feeling of hyper-reality is often felt by participants in edgework, as Lyng found in his work. A skydiver in Lyng's studied claimed that "Free-fall is much more real than everyday existence" (Lyng, 1990, p. 861). Some even find it impossible to find words to describe the experience or even think it is disrespectful to try to describe it.

I find this to be consistent with the experience of technical diving, as time slows down and speeds up depending on the portion of the dive: a few seconds stuck between two rocks in a cave feels like an eternity. Divers often claim to be one with their rebreather, making jokes about sleeping with them. They are connected through the mouth, and the rebreather acts as an extension of the lung—in fact a rebreather is an artificial organ. Divers learn to feel how it reacts and when it is not acting normally. At first, managing a rebreather underwater is overwhelming, and it takes many hours to be comfortable with the machine, just like learning to ride a bike for the first time. When an experienced diver decides to move onto rebreather, he will be told by others that he will be restarting his learning from the beginning because the complexity of the machine makes the new rebreather diver looks like an awkward novice diver. In time though, he will learn to intuitively manage his rebreather by feel. During my deep descent in the treacherous shaft, the feeling of oneness between my mind, my body, and my rebreather exceeded what I had ever felt. The negotiation of the edge by a Deep divers and their attuning can be explained as "continuous entanglements between seeing and moving, prior experiences of seeing or knowing, [that] are always fed back through movement so that [they] are always engaging in the world or medium in new ways. In this notion of perception, it is the active engagement with the environment that elicits stimuli."(Laplante, 2015, p.16) Finally, deep diving by itself brings bliss and solace in a tech-divers life.

Chapter 4. Near miss

During my fieldwork, one of the comments that came back often was that to learn to be a technical diver a novice had to "swallow water" to see what kind of reaction he would have when experiencing a feeling of drowning and a perception of being near death. Such a situation could be considered a limit-experience (Foucault, 1991). In fact, as an extreme sport, technical diving is the venture of a team or individual in a limit-environment, where life is barely sustainable. In such an environment, there are many risks present to divers. But what is a risk? What is a limit-experience, and how does this experience help the development of a technical diver? These are the questions addressed in this chapter.

Risk is part of our daily life: we drive our car to work not knowing if we will get into a collision; we make work decisions that can impact our employment; we sign mortgages with interest rates that may rise resulting in the loss of our house. As I have outlined before, evaluating and negotiating risk is an integral part of technical diving. In order to understand how tech divers perform deep dives, we must understand what risk is.

I begin this chapter with a literature review of existing research on risk in various academic fields. After an overview of risk research in anthropology, I explore the psychological perspective on this topic to outline some of the major theory and differences, which become important in considering diving. Then I summarize the sociological work in the domain of risk-taking and outline the limitation of edgework. Finally, I explore the limit-experience theory of Foucault and discuss near-miss events in diving. The definition of risk takes on many flavors depending on the field of research, so I explain the various definitions where appropriate.

The Anthropology of Risk

The study of risk in anthropology is more recent than in other disciplines, although here I should specify that anthropologists have taken many risks in fieldwork, and all ethnography deals with people and uncertainty at some point since it is an inherent part of human life in any society. Simply put, the definition of risk from the Oxford dictionary is "a situation involving exposure to danger" and "the possibility that something unpleasant or unwelcome will happen." In fact, all societies have dealt with dangers and risks. Many societies have relied on magic, witchcraft, religion, or other mechanisms to explain misfortune when they did not have the knowledge to explain some phenomenon. Mary Douglas's study of witchcraft in *Purity and Danger* originally published in 1966 (Douglas, 2013) allowed her to further understand dangers and risk as well as the symbolism of purity/impurity with the management of threats and misfortune. After understanding that all social groups must face dangers such as natural disasters, she proposed that people can only worry about a finite number of these because of limited time and resources. Trying to attend to and mitigate all dangers would be too time consuming and would keep individuals from functioning in a society. Therefore, individuals select specific dangers to mitigate based on social and political relations unique to that group: those are called risk[23]. Although risk may not be the main focus of anthropology[24], in some of the older work such as the *Argonauts of the Western Pacific* originally published in 1922, Malinowski (2014) questioned the reason why men would risk their lives to cross dangerous oceans to trade what seem to be worthless trinkets. While the main topic of his study, the exchange system of the Kula ring, involved the exchange of gifts linked to political authority, the question of risk was constantly present. In the Trobriand Island,

---

[23] Douglas (2013) explained that in order to properly identify a risk, societies have to place boundaries on the body or the social group. These boundaries are reinforced in a physical way or through a particular belief. When individuals transgress a boundary, there are generally repercussions: in some societies they may receive physical punishment, become stigmatized, or be alienated. By contrast, people who reinforce those boundaries could be glorified or seen as pure, such as the members of a church. Based on Douglas's analysis, the danger would be in and with the person's transition in another state. In that transition, since the state is undefined—the danger would need to be mitigated with a ritual. An individual is therefore inside or outside the social group. When an individual does not conform to the norms and customs of the group, this represents a potential hazard to the order. Therefore, Douglas (2013) argues that it is the loyalty of the individual to the group, balanced with the control of the group over the individual, that determines how risk is perceived.

[24] Anthropologist Deborah Lupton (1999)set out the postmodern position that hazards are socially constructed, which means that they are created by people's decision-making processes. She went on to explain that to a driving instructor, a pedestrian crossing the street would be considered a hazard. This reinforces the fact that risk is socially constructed, and anthropology might be well suited to studying it. Andy Alaszewski (2015) claims that risk logic is shaped by symbolic systems and that those who participate in those systems may not realize what is going on, so an ethnographic study of risk perception is indeed a good way to identify the underlying symbolism. He reinforces the fact that Douglas's research was essential to linking the study of risk with anthropology, as risk was mostly viewed as a subject for psychology-based research.

the karayta'u trade necklaces of shell beads to the north and arm shells to the south, engaging in long, dangerous sea voyages to increase their status and prestige (Malinowski, 2014). The risk for the islanders may be mitigated by a combination of technology and magic, such as large canoes for navigation and Kula magical rituals for protection. Alaszewski (2015) makes the observation that this type of behavior is a form of voluntary risk-taking based on the concept developed by Zinn (2015). Finally, danger has always been present for people, and people have always attempted to rationalize decisions by considering the chances of misfortune. Risk itself might not have been a very popular topic for anthropologist in the past, but people who risk their life for many different reasons have often been at the center of the discipline. It is only natural for anthropologist to pay attention to how people interact with their environment when it inherently involves perceived risk.

Diving and Risk: A Psycho-Mathematical Approach

To my knowledge, there is no anthropological research on technical diving at all, much less on the topic of limit-experience and the perception of risk. However, risk has been studied in other disciplines, including risk connected with diving. The following section outlines some research in psychology[25] about risk and technical diving. Lopes (1987, p. 8) mentions that the word risk refers to "situations in which a decision is made whose consequences depend on the outcomes of future events having known probabilities" and makes reference to common betting games such as roulette. He explains that when an individual does not have knowledge of the probability at stake, the decision is made in ignorance (or uncertainty). Hence, the field of psychology has in the

---

[25] Psychology tends to define risk and danger in its own way and has developed unique theories that can add value to this research. It is also important to understand how psychology is different from anthropology. Psychology attempts to understand the "why," while anthropology seeks to discover "how" things happen.

past studied risk in two different ways: by looking at the decisions an individual makes with regards to gambling from a statistical point of view and by looking at the differences in individuals' personalities in regard to risk-taking[i].

As for technical diving, Gareth Lock recently published *Under Pressure: Diving Deeper with Human Factors*, analyzing various misfortune cases in diving. Lock's book focuses on safety and performance, and he opens one chapter with an anecdote from Steve Bogaert. During a cave diving exploration, Bogaert had laid lines in a remote uncharted section of a cave. During his return, he found himself in a zero-visibility situation where his line marking the way out got buried in the sand, making it impossible for him to follow it and exit the cave. The normal procedure would have been to take out a safety spool, lay lines to bypass the buried line, and find and connect the new line to the exit line to the other side of the obstacle. However, he had lost his safety spool because he clipped it on his rig outside of his pouch so he could carry more equipment and lay more lines. He ended up making a blind jump[26] from one side of the line to the other side in order to save his life (By not connecting the two lines with a spool Bogaert broke one of the most important rule of cave diving; always have a continuous line to the exit). Lock (2019) explains that Bogaert's story reflects the multiple risks and uncertainties that can happen in technical diving where we could face a decision that requires us to choose between two horrible situations such as following the rules by staying where you are on the line and running out of gas, or taking a leap of faith and leaving the rope in a zero-visibility situation in order to attempt to find your way out[27].

---

[26] In penetration, a blind jump happens when a diver moves to a second line (intentionally or by accident) without connecting it to its current line. Because the diver will not have a continuous guideline to the exit, the risk of getting lost increases).

[27] In my opinion, staying on the line and running out of gas was not an option as it would have meant a certain death, so the best odds of making it out alive were by leaving the rope.

Lock (2019) explains that dealing with uncertainty is an important part of technical diving as it is the fear of the unknown that can create panic, which can lead to loss of control and death. He claims that in technical diving we are not managing risk but instead are managing uncertainties, since risks are measurable and uncertainties are not. Risk could be mitigated through logic and data analysis, while uncertainties are informed by emotions and heuristics. To him, engaging in diving means that we cannot remove the probability of death due to drowning[28]. Lock (2019, p. 81) says that "there is no current system to analyze failure and risk therefore many of the risk-based decisions made by divers are mostly considered educated guesses.[29]"

Lock uses technical diving case studies to analyze how humans perform under water and in chapter 4 attempts to answer the question, "why do divers take risks?" He utilizes psychology-based theories for his analysis, and as such I find that he instead answers the question, "how do divers make choices?" since the theoretical background in his book cannot answer why people get into technical diving in the first place. Ultimately, Lock points out that risk or uncertainty can be a good thing since individuals who are risk averse may skip out on opportunities for development. If we could predict everything in diving, this exploration sport would not be as attractive[30]. Finally,

---

[28] Currently rated at 2 deaths per 100,000 active divers (Buzzacott, 2016).

[29] In this statement, I would point out that Lock probably should have used *uncertainty* versus *risk* based on his definition. His statement is only true if adhering to his definition of risk, which is a definition used in project management and finance: the likelihood of an event multiplied by the consequence of the event happening. He then explains the zero-risk illusion to explain that divers tell themselves that "it will not happen to me" since the odds of death are extremely low overall.

[30] Jens Zinn (2015) states that it is counter-intuitive that many people voluntarily take great risk that could result is severe consequences including death, even though health and knowledge have improved in society. Tulloch and Lupton (2003) defined voluntary risk-taking as an individual taking part deliberately in an activity perceived as risky by the participant[30]. Voluntary risk-taking has been identified as an important positive factor in the development of adult identity while growing up; in fact, it is only by taking risks that young people will discover their own limits and gather wisdom. Lyng argued that adults also seek risk to shape their own identities. Although Lyng and Lightfoot do not precise how significant the risk needs to be or what kinds of consequences should be present (participants may have a broken arm versus death) their researches are important as they outline a positive experience provided by taking risk.

Lock's work provides valuable information on and analysis of the human factors involved in technical diving, and I would recommend it to any tech divers. However, looking at how technical divers perceive and take risks in a limit-environment, from an anthropological perspective, provides a much different answer to a similar question answered from a psycho-mathematical perspective.

The Limitations of Edgework

Technical diving fits the description of edgework due to the permanent presence of danger and the possibility of death. The high number of skills involved, the mental focus required to succeed, the feeling of oneness with the rebreather, and the altered feeling of time and space is in line with the concept of edgework. Technical divers often push their limits in order to get as close to the edge as possible without going over it. Although the theory of edgework brings some answers to my question, it fails to explain why a diver would say that a near-miss event is essential to the becoming of a tech diver.

Another point where the concept of edgework does not align with diving is the following. Spontaneity, which is an action that increases the range of human capabilities, first requires constraint, according to Lyng. Of course, the basic humans needs such as food and shelter must be fulfilled. Marx proposed that working in a capitalist condition does not encourage the development of the human capability and intends to reduce humans to a machine in a dehumanizing manner. Edgework would then seem to be a great solution to this problem as a fundamental condition to qualify is the high level of skill required to perform the activity. In turn, it would be this skill-mastery that would lead an individual to think that he could control events that are mostly related to chance, hence providing a great satisfaction from controlling what seems uncontrollable (Lyng,

1990). Lyng (1990, p. 872) claimed that "unquestionably, success in negotiating the edge is, to a large extent, chance determined. Yet edgeworkers generally reject this notion, believing instead that one's survival skills ultimately determine the outcome."

Based on my fieldwork, I agree with Lyng's claim that edgeworkers reject the notion that negotiating the edge is mostly chance determined and instead believe that it is survival skills that determine the outcome. First, if the outcome was largely chance determined, why is there not more death and injury in these sports? Statistically, if the success of a skydiving jump or a technical dive would be mostly chance, the death rate would be excruciatingly high. I agree that there is an element of chance, but this is minimal, with many technologies and skills to respond to those risks.

Second, experienced divers know and understand the risks. Every technical diver has bad stories and made it back alive to tell others the lessons they learned. It is a known fact that a regulator may malfunction and a diver may lose a gas resource during a cave dive, and this is why the training agencies has developed best practices like the rule of thirds, stipulating that divers carry a reserve. Divers who are working the edge and believe that they control all elements of a dive are, in my opinion, inexperienced or immature. The majority of experienced tech divers that I have met acknowledge the level of uncertainty in the dive itself and accept that risk. They know that after a series of malfunctions, or bad luck, they will be out of reserves and will most likely drown. This is why every technical dive that is challenging comes with many hours of preparation, trying to cover most situations that could arise from that "chance" element. Once the planning is complete, the divers must accept the residual risk, or the odds of "having a bad day," which is the expression we use when the only outcome is either a decompression accident or drowning. Those odds are extremely low, and this is why we do not see a 50% fatality rate among technical divers every year. If chance was the biggest factor, there would be more death. Even if skills might be the most important factor for a technical diver's survival, it would be impossible to have an absolute control over the outcome of a dive; there is a minimal residual risk to accept.

Near Miss and Limit-Experience in a Limit-Environment

As seen above, the edgework theory from Lyng does provide some explanation to tech diving, but still has some gaps—not explaining, for example, why many tech divers have claimed that in order to learn, a novice must swallow water. A diver swallows water usually by making a mistake in an out-of-air drill or during a near-drowning experience. During training, it is normal for an instructor to task load the student to a point where he will make a mistake and breath from a depressurized, turned off cylinder, which results in breathing in water. In a training setting without a decompression obligation or ceiling, this will usually surprise the student and sometimes create a feeling of anxiety, but without serious threat to his or her life. However, when a diver loses himself in a cave and is about to run out of air or swallows water thinking he is out of air and there is an imminent threat to life, then it produces a unique feeling. Such an event can be classified as a near miss[31]. The following story illustrates a near-miss event.

Lost in a cave, Sheck Excley's regulator started to breathe hard as a sign that his air reserve was running out quickly. Because of technical difficulties related to his reel in a previous dive, he chose not to run a line in an exploration dive, making it impossible to locate the exit. While panic was threatening his life and making him unable to decide which tunnel to pick as an attempt to save his life in this underwater maze, he finally calmed down in acceptation of death: "Somehow

---

[31] The term near miss is also used by Wade Davis (2011)to describe a situation when a serious accident or a disaster very nearly happens during the conquest of Everest.

my mind cleared, perhaps from a blind surge of determination to get out but more probably from the calmness accompanying resignation to death" (Exley, 1994, p. 28).

Figure 25 - Train track leading to the exit in Forsyth Mine (Photo by Kevin Brown)

This situation is what I call a near miss, a term used in technical diving to represent an event where a participant had a misfortune and faced an imminent threat to his or her life. Foucault's explanation of a limit-experience can help clarify this concept. I believe that all near-miss events are limit-experiences, although not all limit-experiences are near misses. A near-miss event implies that the person experiencing it believes that he or she will not survive. In a way, it is a limit-experience where divers perceive themselves going over the limit.

In my early 20s, I would visit North Florida at least twice a year to go cave diving. A few trips after I got my full cave qualification, I started to dive sidemount with my own custom rig as the commercial options were still limited (the armadillo being the most common). There was no sidemount course so I had to learn by myself, and this was a somewhat natural progression from a standard twin set. One of the local cave systems, very popular at the time, offered the perfect opportunity to push one's sidemount skills in area that had already been lined. I always felt more secure diving solo for this type of dive as I thought that another diver would just double the chance of the exit being blocked in a bad situation.

I started the dive with my mentality of invulnerability, talking a jump into a very small and narrow tunnel going down to about 120 ft of depth, not particularly deep for me now, but at the time in open circuit a depth that would make my air run out quickly. As I was penetrating the cave, I realized that the line was unusually small, eroded, and dangling, meaning that I was in an area that was not commonly visited by cavers. I started to feel the coffin effect—moving forward in a tight tunnel without enough space to turn around to exit feels like being stuck in a coffin. Moving backward was not a great option as it would take at least twice the amount of air since it would be slow and laborious. At one point there was enough space to bend myself in half and turn around. I then followed the line in zero visibility due to a few restrictions that I had to go through combined with the fact that there was no flow in this area of the cave. I came to a point where I was completely wedged between the ceiling and the floor, with the line going in a direction where the diameter of the cave was so thin that I knew I could not have passed there to come in.

Stress began to build. Was I lost and following a different line? Did I accidentally jump during this blind exit? I came to realize that this was a line trap: the dangling line had slipped to a different position, making it impossible to keep contact with it and pass the restriction. Paralyzed between two rocks and blinded by the silt, I came to realize that my air was running out quickly. Stress and panic would only make things worse by increasing my rate of breathing. I knew I had to make a decision, since the status quo would lead only to my death. With my air reserve already low, I thought to myself that this was going to be the end, this would be my coffin. This acceptance of death did not lead to panic, but instead to calmness. Only by shifting my mind to a "dead" state could I get back alive, bringing hope instead. The calmness of being in a place that I chose to be in doing what I loved was enough to accept this death, slowing my breathing and sharpening my focus. I could tie a safety spool into an already loose line to try to find another way, let go of the rope or just simply pull on it risking breaking the old and delicate line. Having one loose string already threatening to get tangled around my equipment was enough, so I decided to commit to a leap of faith and pulled the rope to search for the restriction in this zero-visibility situation. Minutes felt like hours, but in the end I came out alive, eating through my last third of gas. This was my first near miss in diving.

Was I ever in real trouble or was it only my mind playing tricks on me? I had the skills to pull it off but lacked experience, though this conclusion is only possible when looking back at an event with my current experience. My experience now would dictate a whole different approach to this kind of dive and I would certainly bring many new tools to the game, therefore the same situation would probably not result in a near miss again. At the time though, this was how I felt. The first near miss that I ever experienced taught me many things, including that I was not as invincible as I thought I was. It also made me question why I was doing this sport, as it seemed like there was nothing tangible at the end of this tunnel, especially since it had been previously explored. Ultimately, I believe it was a growth experience, quite inevitable in such a sport.

Andrew also had a near-miss experience. Exhausted after a dive, he was unable to re-board the dive boat due to extreme weather. While getting rocked at the surface, he believed that he was going to faint and not make it out alive. In a similar situation, Roger suffered a CO_2 hit trying to fight waves and currents at the surface, trying to reach the boat. As he started to have tunnel vision, a thought of death came to his mind. All of those near-miss situations share similarities with a limit-experience; they tear the diver from himself in a way that he will not be the same after the event. By living through an intense feeling at the limit of life, divers grow. Although it is perceived as a negative event in the present, in the future it is often described as an event that is "quite essential to learn" (Andrew). During a discussion with Roger in the days following his incident, he told me that he needed those kinds of situations every now and then to remind him that he is alive and make him feel alive. In a way, he enjoys those near misses. Just like a limit-experience, a near-miss turns from negative to positive—a principle that Foucault learned from Georges Bataille.

Where I believe that near-miss events differ from limit-experiences is that a subject cannot voluntarily and deliberately experience a near-miss event; it must be a surprise so that the diver, at the time, will not think that he will survive. Of course, if the diver does not survive, then it is simply a "miss." In a way, it is a limit-experience with the diver's perception of "beyond limit-experience". A very challenging deep dive in a mine could be considered a limit-experience; when the diver's regulator stops delivering gas unexpectedly, then it could be a near-miss.

A near-miss event can only be resolved by the acknowledgement of death. The other option, panic, would most likely end up in a fatality. Diving is mostly a binary sport with little in-between—a diver either succeeds or does not, breathes or does not. In fact, there have been very few successful cave diving rescues, as opposed to body-recovery missions. I did participate in a few such missions along with other technical divers in Outaouais. As an experienced technical diver in Quebec I felt I had the responsibility to partake in the searches as the police department's capability is too limited in terms of depth. During a dive, there is almost no likelihood of broken bones due to falling or other common injury, although I concede that some people do survive a bad decompression sickness and remain crippled. The near miss is when the mind alternates between the binary state of diving, switching from alive to dead to alive. I could almost compare a near miss to Schrodinger's Cat, where in this experiment the cat is considered both alive and dead as he is in a state known as a quantum superposition.

Many divers have during their careers experienced a near-miss event, where they have faced death, acknowledged it, and worked their way out. It is that split second where the mind realizes that this could be a no-exit scenario, where time is distorted to make it feel like hours instead of seconds, which is crucial to the survival of the diver. One can take those newly created precious hours to ponder happy memories of family, successes, or failures, or to focus on the task at hand. Surely, focusing on memories will ensure that a diver's life ends in a happy state while doing a sport that he or she enjoyed versus in the mode of panic, fear, or pain. The acknowledgement of death makes one, in a way, invulnerable. The determination is shifted to coming back alive from a state of happiness where time is no longer an issue. In fact, all other issues, problems, and trauma in life have magically disappeared; the pain gives way to joy. It is a special moment of pure freedom, where any other life problems disappear. It is resilience that brings the diver back out of the water. A near miss is such an intense event that it must be decompressed, ether underwater or as a social event at the surface.

But the trap is set: ponder too long in that happy place and a diver risks never returning, transforming the "near-miss" event into a pure "miss." The journey to the other side is a quest to understand and remember what one lives for, and that ultimately to remember that one *is* alive. In a post-near–miss incident, Andrew would describe the moment as "you know that you are alive," while Roger would say, "it reminds you that you are alive." When divers share near-miss events or limit-experiences, it makes them grow. They either develop trust that becomes the foundation of the dive team or they realize that their partner is not trustworthy. How can divers deliberately go through a limit-experience if they do not believe their partners will help them in times of need?

Limit-Experience and Deep Diving

I have already established that diving is mostly binary. Therefore, the ultimate risk we need to manage is death itself. Before a deep dive, after a diver has done everything in his power to lower his risk, he manages the residual risk by accepting it, by accepting that if something goes wrong, it could mean death. A deep diver makes peace with it. He writes his will. He ensures that his life is in order. By being forced to accept this residual risk to perform a deep dive, I put order into my life and keep it on track in case the dive would go wrong. As opposed to edgework, I do not see this kind of deep diving as a way to control a situation that seems uncontrollable. I know there is much out of my control in this dive, and I accept the uncontrollable event.

I consider depth with ceilings to be a limit-environment. Of course, what depth and conditions constitute a limit-environment depends on the perception and ability of the diver. A pool is not a limit-environment, although many humans have drowned in one. It is an environment that barely sustain life. An environment like a deep cave or a deep mine will have a zone where life is easily sustainable such as near the surface. It may as also have a zone where human life is prohibited even with technical diving equipment. The limit-environment will be in between, as close to the limit of survival as possible.

Divers crossing the line beyond edgework, trying to push the limit-experience, do accept a high level of risk and the number of deaths beyond the edge is incredibly high. But these divers do not lie to themselves; they understand that the odds of dying might be 25% to 50%, perhaps higher.

114

The illusion of control over the environment also does not apply here, since the divers are experienced enough to understand the odds and they are also surrounded by a team that will help them achieve their goals—that team should remind the tech diver of the risk, and significant discussion will arise.

For example, during the exploration of the Forsyth Mine at 602 ft in freezing water, I estimated the odds of not making it out at around one in three. I was going to be the first one to achieve this depth in the mine where we had no idea how stable it was, and my worst fear was that the bubble created by my rebreather during the ascent in the elevator shaft would create great pressure on the ceiling, making it collapse and trapping me forever. This was a risk that I could not control, and I knew there was nothing to be done. "Do it, or don't do it—that's it," I told myself. There were also other risks that I could not plan for, such as piercing my suit and flooding with frozen water while having many hours of decompression left. I knew that the mine had many sharp objects such has nails and rusted metal pipes, and that in order to exit there was a nasty restriction around 90 ft, so that if I tore my suit I was done. This was another risk that contributed to the math. In my diving career, I knew how many times I had pierced my suit in a wreck: only twice in over 20 years of diving. I therefore knew that this risk was small enough for me to accept it. I had numerous discussions with other tech divers, and there was a consensus that the odds of not coming out of such a dive alive were not trivial. That is why it was not something I repeated every weekend. There was no illusion of control over this dive; I was well aware of the odds regarding the outcome, just like the astronaut landing on the moon knew that there was a risk of non-return.

Just like the Trobriand islanders risked their lives to exchange gifts with the aim of gaining social recognition in Malinowski's ethnography, tech divers take great risk venturing to depth as a demonstration of skill, courage, and mastery, in return gaining social reconnaissance of other divers, even though the achievement usually means nothing to the rest of society, the non-divers. To be human is to challenge limits, to learn, and to grow. Accepting a challenge that is both physical and psychological with a significant chance of dying is pushing the limit of the limit-experience and challenging death itself. Redefining the limit of my death zone is not only critical to who I am, it is also where I found what is important in my life itself. That limit-environment filled with adrenaline, silence, dark, and cold is where I find freedom and peace. A diver will discover a great deal about himself during the journey to the depths, where the rigid rules of society do not apply.

In summary, the process of becoming a technical diver is a lifelong journey where a diver learns to adapt to an environment hostile to human life. The basic skills are acquired during classes to ensure that a novice diver will survive in this limit-environment. By being part of a small team, a diver will gain knowledge passed on from more experienced members. As he takes more risks to go deeper and longer, he will go through a series of limit-experiences and near misses that are essential to his development. Risk, as the perception of danger, is often increased proportionally to the decompression requirement and time under ceiling. In turn, those limit-experiences and near-miss events shared with teammates create mutual trust. The sharing of difficult and painful experiences makes people understand how they will react and how their team members will react during such times. It is this trust that becomes the foundation of the team and allows the team to improve upon existing techniques and increase the depth and difficulty of their dives. In order to accept great risks, divers need to trust that their team members will be there for them. After a dive, divers will decompress and further increase the social bonds within the team. As generalization is never the aim of anthropology, I would like to specify that near-miss events are surely lived and perceived differently by other people; this is only one story.

Conclusion – Decompression

In closing, the process of becoming a deep diver is a lifelong journey where experience and wisdom are developed through multiple dives composed of compression-decompression cycles.

Throughout this book, I described my own experience through the school of diving, where I learned the basic of technical diving. The classroom and practical drill exercises teach the novice how to survive underwater, but the classes are rather short in duration and limit the exposure of the novice. To become a deep diver, a tech diver needs to learn how to bend those set of rules and standards to adjust for the ever changing environment.

Depth itself is a finality. Just like the dive at Forsyth mine, the planning sometimes begins a year before its execution. During the execution, the physical compression of the human tissues takes place during the descent and time spent at the bottom of the immersion. Risk is important as it is the perception of danger in an environment. It is this perception of great danger that creates the qualification for a *milieu limite* where edgework is practice. Trimix, a mix of helium, oxygen, and nitrogen allows divers to venture deeper into limit-environments such as mines, wrecks, or caves. Those environments are particularly attractive to technical divers as they provide an overhead environment, increasing the difficulty, technicality, and risks. Those perceived risks make such environments appealing to these divers; and above all is the excitement of laying lines in an unchartered maze. Deep diving with mixed gas comes with the requirement for decompression, where time slows down and the diver's tissues can purge the excess inert gas. An omission of the decompression schedule, or simply a bad day, can trigger decompression sickness, also called the bends. Those bends can hurt, cripple, or kill a diver depending on the severity.

Perhaps technical diving is best described and represented by the need to solve problems at depth, since getting out of the water in the face of danger is not an option. The stories about deep diving presented in this book provide insight into what it feels like to be diving and what goes on in a diver's mind during a dive. However, stories cannot replace the experience of deep diving itself. I hope that these stories were sufficient to demonstrate that these kinds of dives executed in cold water required flexibility from divers; it required to bend the rules. By challenging the assumptions of the past, a team of technical divers constantly innovated and changed processes and procedures to push the limits of the sport.

Total mastery of diving is never really achieved, as the process of becoming a deep diver is a lifelong process. Many novices see me as quite experienced. However, with only twenty years of experience in the sport, I am only starting to get it right; the more I dive, the more I discover that I know little. To attune to a hostile environment, divers dedicate much time and resources preparing and training, facing multiple risks. Nonetheless, their multiple limit-experiences shared with their team are critical to developing the trust that is essential to pushing limits even further. Within a team, knowledge, experiences, and risks are shared. While pushing limits to the edge, divers experience near miss events, unforecasted incidents where divers believe that they will die. During those events, divers momentarily cross the line of the limit-experience and face death, providing them an ultimate learning experience. Near miss experiences are part of the learning process and have been deemed an essential part of becoming: you cannot learn without swallowing water. In situations where divers face death they will learn about themselves; will they succumbs to panic and drown, leave a team member behind in a difficult situation, or attempt a rescue? By surviving such situations, tech divers grow together and develop an extremely strong bond based on trust, which is the foundation of the dive team. At depth, on the edge, the diver will find bliss and solace. Therefore, near misses are deemed regenerative.

Finally, the decompression phase begins with the ascent of the diver from the maximum depth. At first we see the physical decompression of the human tissues, which usually continues for more than 24 hours after the dive. In the shallower and lengthy decompression stops, where the risk of the dive itself often appears to be reduced and adrenaline dissipates, the diver's mind begins decompressing. This could also explain the importance of the social aspect in the post-dive activity, such as drinking a cold beer with a warm soup at the local pub amongst friend. Social decompression of near-miss events is mandatory for the learning experience to be positive. I would hope that the diving term near-miss could be used in anthropology to fill some of the gap of the limit-experience.

There are many ways to dive, and this book does not apply to everyone everywhere. It is merely a window into the life of a few technical divers. In closing, tech divers often put their lives at risk, and tech diving may seem like an unreasonable sport to practice since many people have died doing it. I have been diving for a long time and have had my share of incidents. This book marks only the beginning of a long journey of studying technical divers, and I hope it will bring a better understanding of what it feels like to be alive at depth. The presence of lines for instance emerges implicitly throughout the text; when I talk about bending the rules, when I describe the edge of edgework, or when discussing the limit of the diving technique. In retrospect, the chapter of the book a Thousand Plateaus from Gilles Deleuze and Felix Guattari (1989, p.208-p.231),

Micropolitics and Segmentarity, seems to provide a theoretical toolset appropriate to comprehend the complexity of diving. For example, some organizations are molar, such as the regulatory agencies, while others are molecular, flexible. While underwater, some movements are extremely controlled (molar) when negotiating the edge. The near-miss experience could be seen as a line of flight completely uncontrolled, while quite essential to the becoming of a deep diver, offering other possibilities for further work in anthropology and technical diving, routes already explored by Ingold (2015) that I could be particularly fruitful for this topic.

Further study is necessary to understand aspects like the affections that tech divers share for objects such as tea cups that they collect in shipwreck, and why they risk their lives and sometimes perish for such economically worthless objects. The attraction of shipwrecks to tech divers and their relationships with underwater archaeologists should be further investigated.

[i] In the mathematical realm, experimental psychologists such as Kahneman and Tversky (1979) have concluded that most individuals would rather take $3,000 now than have an 80% chance of gaining $4,000, even though the expected value of the second choice, at $3,200, is higher. They have explained this adversity to risk by claiming that people prefer to exchange potential return with risk (Lopes, 1987), which is exactly how many banks present a type of portfolio available for investment: a lower risk will yield lower return. Another way to explain the decision of choosing a lower expected value in a game was set out by the famous mathematician Bernoulli in the 1700s (Bernoulli, 1954) who understood that money was not proportional to its utility. The value of money depended on the amounts of wealth and goods already possessed by the individual facing the choice. For example, a poor person who cannot feed his family today would be more likely to take a certain $1,000 than a 50% chance of $3,000 as opposed to a rich person who has million in the bank who is most likely to choose the higher expected value. This is because the $1000 has more utility to people who have lower resources, hence the value of money is not linear. This phenomenon could be explained by Maslow's (1954) pyramid depicting the hierarchy of need stating that someone physiological needs require to be satisfy before the self-actualization needs. Finally, psychologists who have studied risk and choice have found that people often replace the actual probability of success with a subjective one (such as winning the lottery) or modify the value by adding an artificial weight. Psychologists have also studied risk and decision-making from a personality perspective by looking at motivation and incentives.

Lopez (1987) has examined the famous "ring toss" game used by David McClelland, which was to throw a ring into a peg where the player could choose anywhere between a very short and extremely long distance. The result was that people with high motivation usually chose a moderate distance that had a reasonable probability of success but was still difficult enough to provide an achievement incentive, as opposed to people with lower motivation to succeed who would either stand very close to the peg or extremely far away, making the shot a sure or impossible one and thereby removing the performance anxiety.

In *Between Hope and Fear: a Psychology of Risk*, Lopes (1987) presented the two factor theory by combining the strength of both psychophysical and motivational theories aiming at explaining the hopes and fears that give an experiential texture to risky choices. She defines the term "risky choice" as a decision that incorporates an element of danger and is different from a simple win or lose gamble, which has been the focus of most research. Her research found that a single individual may demonstrate risk-averse choices and risk-seeking choices. In fact, risk aversion is the most common behavior (Pratt, 1964), and this is the natural survival instinct passed down to younger generations: for example, the safety-first principle among subsistence farmers is to plant crops to feed their family first and then plant seeds that are higher value for selling. Because of that principle, taking risk can be seen as a luxury: one needs enough basic wealth to allow oneself to take more risk. Risk-seeking people would be motivated by the desire for potential or by being in a bad position that forces them into taking more risk because there is no safer option (Lopes, 1987). To sum up, psychology has focused on risk behavior as a negative aspect that harms individuals when they fail to evaluate the risk or make the right choice based on mathematical assessment and illusions or based on motivation theory; the ultimate aim is to diminish risk for the best expected value. This psychology approach might not be particularly attractive to the anthropologist, since it cannot explain the reasons why some people enjoy taking risks or why they would voluntarily venture into a limit-environment and risk their own lives.